Classroom Companion: Business

The Classroom Companion series in Business features foundational and introductory books aimed at students to learn the core concepts, fundamental methods, theories and tools of the subject. The books offer a firm foundation for students preparing to move towards advanced learning. Each book follows a clear didactic structure and presents easy adoption opportunities for lecturers.

Dmitry Ivanov

# Introduction to Supply Chain Analytics

With Examples in AnyLogic and anyLogistix Software

Springer

Dmitry Ivanov
Department of Business and Economics
Berlin School of Economics and Law
Berlin, Berlin, Germany

ISSN 2662-2866 ISSN 2662-2874 (electronic)
Classroom Companion: Business
ISBN 978-3-031-51240-7 ISBN 978-3-031-51241-4 (eBook)
https://doi.org/10.1007/978-3-031-51241-4



This Springer imprint is published by the registered company Springer Nature Switzerland AG
The registered company address is: Gewerbestrasse 11, 6330 Cham, Switzerland

# Preface

Everyone makes decisions. In industry and services, supply chain managers take decisions facing different questions, e.g.:

- What are the best locations for our warehouses, distribution centers (DC), and production sites?
- How to allocate demand and supply in a complex network?
- What are the best policies for replenishment, sourcing, and transportation?
- How resilient is our supply chain to supplier disruptions?
- Why does our on-time delivery performance decrease when we change our inventory policy?
- What will happen with demand fulfillment rate if we change a DC's capacity?
- How can we use sales and user experience data to maximize revenues by price adjustments?
- Do we have enough inventory and production capacity if demand would suddenly increase?

Decision can be made by intuition or based on facts, historical data, and model-based prescriptions and forecasts. Data-driven, model-based decision-making is a subject of supply chain analytics. Supply chain analytics is a topic of growing interest for academia and practitioners alike. Despite the increasing interest in and importance of supply chain analytics, there is no concise and generally understandable source which could be taken as an introduction to supply chain analytics.

This book offers a concise yet comprehensive introduction to supply chain analytics covering management, modeling, and technology perspectives. Designed to accompany the textbook "Global Supply Chain and Operations Management," it addresses the topics of supply chain analytics in more depth. The book describes descriptive, predictive, and prescriptive supply chain analytics explaining methodologies, illustrating method applications with the use of training exercises, and providing numerous examples in AnyLogic and anyLogistix software. Throughout the book, numerous practical examples and short case studies are given to illustrate theoretical concepts. Along with AnyLogic and anyLogistix model development guidelines and examples, the book has two other distinct features. First, it reviews and explains novel frameworks and concepts related to data-driven decision-making and digital twins. Second, it shows how to use analytics to improve supply chain resilience.

Without relying heavily on mathematical derivations, the book offers a structured presentation and explanation of major supply chain analytics techniques and principles in a simple, predictable format to make it easy to understand for students and professionals with both management and engineering background. Graduate/PhD students and supply chain professionals alike would benefit from a structured and didactically oriented concise presentation of the concepts, principles, and methods of supply chain analytics. Providing graduate students and supply chain managers with working knowledge and tools for basic and advanced supply chain analytics, this book contributes to increasing knowledge awareness of decision-making in data-driven and digital environments. The book is supplemented by a companion website offering exercises with the use of AnyLogic and anyLogistix software as well as spreadsheet modeling.

The author wishes to thank Mrs. Jialin Yin, Associate Editors at Springer Nature, and the entire Springer production team for their assistance and guidance in successfully completing this book. I also thank all colleagues working in supply chain analytics area for publishing their works and sharing the knowledge. I thank my research associates, Phu Nguyen and Hiran Prathapage, for their great assistance in proofreading and visualization. Last but not least, I cordially thank my family who supported me enormously during our work on the book.

**Dmitry Ivanov**
Berlin, Germany
September 2023

# Contents

## Supplementary Information

# About the Author

**Dmitry Ivanov**
is Professor of Supply Chain and Operations Management at Berlin School of Economics and Law (HWR Berlin) and Faculty Director of M.A. Global Supply Chain and Operations Management program at HWR Berlin.

His *research* explores structural dynamics and control in complex networks, with applications to supply chain resilience, Industry 4.0, supply chain simulation, risk analytics, and digital supply chain twins. He is co-author of structural dynamics control method. His research coined several seminal academic and practical directions such as the ripple effect in supply chains and supply chain viability. He has applied different methodologies such as optimization, simulation, control theory, and artificial intelligence to a variety of supply chain and operations management problems. Most of his research stems from real practical context and focuses on the interface of supply chain management, operations research, industrial engineering, and digital technology.

He has been *teaching* courses in operations management, supply chain management, logistics, management information systems, and strategic management at undergraduate, master's, PhD, and executive MBA levels at different universities worldwide in English, German, and Russian for about 25 years. He has given guest lectures and webinars, has presented scholarly papers, and has been a visiting professor at numerous universities across the globe. He is the author of anyLogistix educational complex for supply chain optimization and simulation that is in use at more than 300 universities worldwide. He enjoys being part of students' learning experience and creating a safe and active learning environment. He sees his job as an educator to equip future industry leaders with working knowledge and skills in management and technology that would shape more resilient, adaptable, and sustainable supply chains and operations.

His *academic* background includes industrial engineering and management, operations research, and applied control theory. He studied industrial engineering and production management and graduated with honors. He has gained PhD (Dr. rer. pol.), Doctor of Science (ScD), and Habilitation (Dr. habil.) degrees. Prior to becoming an academic, he was mainly engaged in *industry and consulting*, especially for process optimization in manufacturing and logistics and ERP systems. His practical expertise includes numerous projects on the application of operations research and process optimization methods to operations design, logistics, scheduling, and supply chain management. Prior to joining the Berlin School of Economics and Law, he was Professor and Interim Chair of Operations Management at the University of Hamburg.

Professor Ivanov's *research* makes an impact. His papers belong to the category of most cited worldwide in the areas of supply chain resilience and digital supply chain. For example, he has been the most cited author of premier journals like International Journal of Production Research and Transportation Research Part E in 2020–2023. His research record includes over 420 publications, with over 150 papers in prestigious academic journals, three editions of the leading textbook "Global Supply Chain and Operations Management," "Introduction to Supply Chain Resilience" book, and several research books such as "The Digital Supply Chain," "Supply Networks Dynamics and Control," "Structural Dynamics and Resilience in Supply Chain Risk Management," "Scheduling in Industry 4.0 and Cloud Manufacturing," and "Handbook of Ripple Effects in the Supply Chain." Professor Ivanov's research has been published in various academic journals, including Annals of Operations Research, Annual Reviews in Control, Central European Journal of Operations Research, Computers and Industrial Engineering, European Journal of Operational Research, Expert Systems with Applications, IEEE Transactions on Engineering Management, IISE Transactions, International Journal of Information Management, International Journal of Integrated Supply Management, International Journal of Inventory Research,

International Journal of Physical Distribution & Logistics Management, International Journal of Production Research, International Journal of Production Economics, International Journal of Technology Management, International Journal of Systems Science, International Transactions in Operational Research, Journal of Scheduling, Omega, Production Planning and Control, Transportation Research Part E, etc.

He is a recipient of IISE Transactions Best Paper Award 2021, Best Paper and Most Cited Paper Awards of International Journal of Production Research (2018, 2019, 2020, 2021, 2022, 2023), Omega Best Paper Award 2022, Annual Reviewer Award of International Journal of Production Economics (2020), Commended Paper Award at International Conference LogDynamics (2018), and Clarivate Highly Cited Researcher Award (2021, 2022, 2023). He is listed in WiWo ranking 2018, 2020, and 2022 "The Best Researchers in Business and Management" in the category TOP 50 (ranked #5 in 2022). His research projects have been supported by funding from the European Commission (Horizon 2020), DFG (German Research Foundation), IFAF Berlin, as well as industrial companies. He has been the principal investigator in several research projects on resilience and digital twins, including European projects ACCURATE and CERERE (2023–2027) and the DFG Collaborative Research Cluster on Resilience of Global Supply Chains at HWR Berlin (2023–2029).

Dr. Ivanov is profiled in supply chain and operations management, operations research, and industrial engineering. He is passionate about bridging the knowledge of different disciplines and applying it to the solution of practically relevant problems. He has delivered numerous invited plenary, keynote, and panel talks at the conferences and global webinars of INFORMS, POMS, IFAC, IFIP, IFPR, DSI, and Academy of Management. He is leading working groups, tracks, and sessions on the digital supply chain, supply chain risk management, and resilience in global research communities. He is Editor of International Journal of Integrated Supply Management and an Associate Editor in International Journal of Production Research, International Transactions in Operational Research, International Journal of Systems Science, and Omega. He is an editorial board member and associate and guest editor in different journals, including Annals of Operations Research, Annual Reviews in Control, IISE Transactions, International Journal of Production Economics, International Journal of Production Research, International Transactions in Operational Research, International Journal of Information Management, International Journal of Inventory Research, International Journal of Physical Distribution and Logistics Management, International Journal of Systems Science, and Omega. He is Chairman of the IFAC CC5 "Cyber-Physical Manufacturing Enterprise," former Chairman of the IFAC TC 5.2 "Manufacturing Modelling for Management and Control" (2017–2023), and Co-Chairman of the IFAC TC 5.2 Working Group "Supply Network Engineering." He regularly presents his research results and has been Chairman, IPC, and Advisory Board member of over 70 international conferences in supply chain and operations management, industrial engineering, control, and information sciences where he has organized numerous tracks and sessions (including IFAC MIM, IFAC INCOM, EURO, INFORMS, IISE, IFORS, DSI, POMS, OR, IFAC World Congress, IFIP PRO-VE, and LDIC, to name a few). He has been General Conference Chair of IFAC MIM 2019—one of the worldwide largest conferences in manufacturing, industrial engineering, operations, and supply chain management (750 participants).

# Abbreviations

| | |
|---|---|
| **AI** | Artificial intelligence |
| **DC** | Distribution center |
| **ELT** | Expected lead time |
| **EOQ** | Economic order quantity |
| **ERP** | Enterprise resource planning |
| **FIFO** | First in, first out |
| **FTL** | Full truckload |
| **GFA** | Greenfield analysis |
| **HILF** | High-impact-low-frequency |
| **KPI** | Key performance indicator |
| **LIFO** | Last in, first out |
| **LIHF** | Low-impact-high frequency |
| **LTL** | Less than truckload |
| **MAD** | Mean absolute deviation |
| **MAPE** | Mean absolute percentage error |
| **MSE** | Mean squared error |
| **NO** | Network optimization |
| **PLE** | Personal learning edition |
| **RFID** | Radio-frequency identification |
| **ROP** | Reorder point |

# Analytics and Model-Based Decision-Making Support

Contents




D. Ivanov, *Introduction to Supply Chain Analytics*, Classroom Companion: Business,
https://doi.org/10.1007/978-3-031-51241-4_1

1

**Learning Objectives**

In this chapter, we introduce major terminology, methods, and applications of supply chain analytics. The chapter starts with fundamental definitions of supply chain management, data, models, systems, and data-driven, model-based decision-making support. Subsequently, classifications of supply chain analytics methods are presented explaining principles of optimization, simulation, and artificial intelligence. A special focus is directed towards descriptive, predictive, and prescriptive applications of analytics in supply chains.

To this end, our learning objectives for this chapter are as follows:

- Major decisions in supply chain management.
- Principles of data-driven, model-based decision-making support.
- Supply chain analytics methods.
- Descriptive, predictive, and prescriptive applications of analytics in supply chains.

## 1.1 Basics of Supply Chain Management

### 1.1.1 Supply Chain Management

A *supply chain* is a network of organizations and processes wherein a number of various enterprises (suppliers, manufacturers, distributors, and retailers) collaborate (cooperate and coordinate) along the entire value chain to acquire raw materials, to convert these raw materials into specified final products, and to deliver these final products to customers (see ◘ Fig. 1.1).

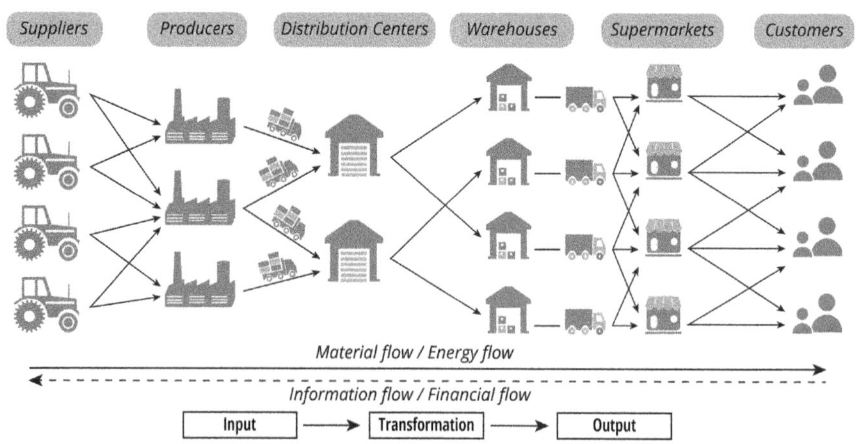


◘ **Fig. 1.1** Supply chain

*Supply chain management* (SCM) is a "cross-department and cross-enterprise integration and coordination of material, information, and financial flows to transform and use the supply chain resources in the most rational way along the entire value chain, from raw material suppliers to customers."

SCM is one of the key components of any organization and is responsible for balancing demand and supply along the entire value-adding chain (Ivanov et al., 2021a, p. 9). SCM, as the term implies, is primarily focused on the interorganizational level. SCM integrates production and logistics processes. In practice, production, logistics, and SCM problems interact with each other and are tightly interlinked.

## 1.1.2 Decisions in Supply Chain Management

The main management task is making decisions. Decisions in SCM are primarily directed towards matching demand and supply by designing and operating the transformation process in the most efficient way. Put simply: SCM is building a bridge between customers and suppliers (◘ Fig. 1.2).

Consider an example. A chocolate *supply chain* can be used as an illustration for depicting basic decisions in SCM. What happens between the cocoa pod harvest and the placement of the chocolate on supermarket shelves? A complex supply chain is built to connect raw material suppliers and end customers (see ◘ Fig. 1.3).

To produce chocolate, cocoa pods are first harvested from cocoa trees, e.g., in Côte D'Ivoire. Cocoa pods are then moved by donkeys to a processing station where they are packed into special carrier bags to avoid damages during transportation by container ship. At the harbor, the bags are packed into special containers and moved to a container ship that will bring them, e.g., to Hamburg.

◘ **Fig. 1.2** SCM as a bridge function for matching demand and supply

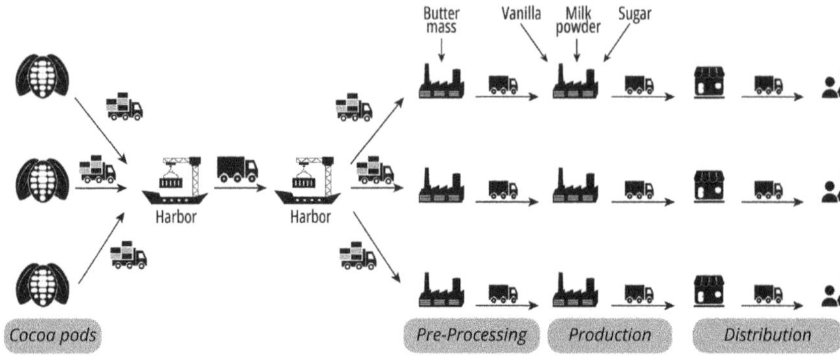


■ **Fig. 1.3** Chocolate supply chain

Along with sugar and milk powder, vanilla is also one of the components needed for chocolate production. Sugar and milk powder are usually sourced locally to avoid long transportation. After unloading at the Hamburg harbor, through which up to 200,000 tons of cocoa pods are shipped each year, the transportation is continued by trucks. Simultaneously, container ships in Guatemala and Madagascar leave the harbor with cargoes full of vanilla.

The cocoa pods are delivered to a preliminary processing plant and separated into cocoa butter and cocoa mass, which are then moved in trucks by road to chocolate manufacturers. After getting all the ingredients, a multistage manufacturing process is started, the final result of which is chocolate. Chocolate is then packed and delivered in large batches on pallets to distribution centers (DCs). From here, small batches are finally delivered to supermarket.

It can be observed from the chocolate supply chain that the responsibilities of SCM can be divided into sourcing, production, distribution, and transportation. The decision-making areas range from strategic to tactical and operative levels (■ Fig. 1.4). *Strategic* decisions include, for example, optimal locations or DCs, decisions on the structure of service networks, factory planning, and design of the supply chain. *Tactical* decisions are production, transportation, and inventory planning. *Operative* decisions are comprised of production scheduling and control, inventory control, quality control and inspection, vehicle routing, traffic and material handling, and equipment maintenance policies.

This description holds true for many different organizations including global brand manufacturers, major retailers, nonprofit organizations, and local petrol stations or hospitals. Purchasing, assembly, shipping, stocking, and even communicating are a few examples of the many different actions unfolding within these organizations, all united by a single purpose: to create value for a customer. Moreover, SCM methods are of vital importance to design and run firms and supply chains in a resilient manner (i.e., to be able to cope with disruptions and recover). One example is the COVID-19 pandemic that required unprecedented efforts from supply chain and operations managers.

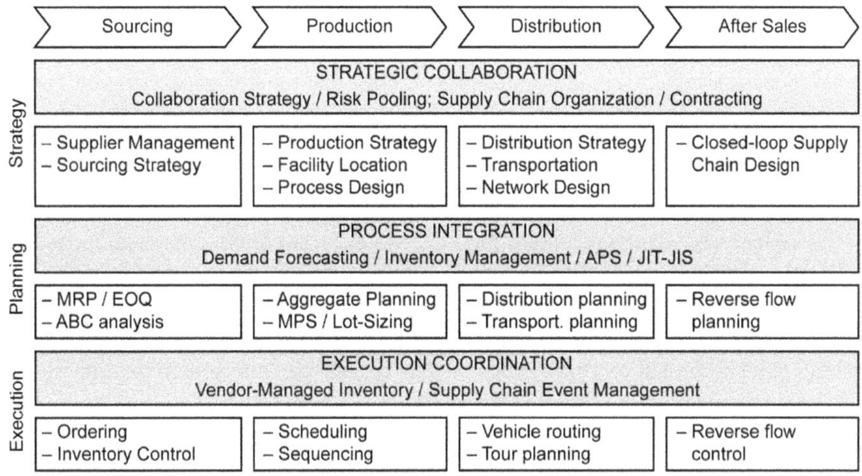


**Fig. 1.4** Decision matrix in SCM (Ivanov et al., 2021a)

## 1.2 Problems, Decisions, Systems, Data, and Models

### 1.2.1 Problem and Decision

For a *problem* to exist, there must be an individual (or a group of individuals). This person is referred to as the problem owner(s) (decision maker) who:

- Is dissatisfied with the current state of affairs within a real-life context or has some unsatisfied present or future needs, i.e., who has some goals to be achieved or targets to be met
- Is capable of judging when these goals, objectives, or targets have been met to a satisfactory degree
- Has control over some aspects of the problem situation that affect the extent to which the goals, objectives, or targets can be achieved (Daellenbach & McNickle, 2005)

Rather than assuming that we have a well-structured problem with clearly defined objectives and alternative courses of action, it is better to:

- Represent the problem situation where issues are still vague or fuzzy
- Gain a comprehensive understanding of the various issues involved in decision-making
- Formulate the problem correctly to ensure an appropriate level of detail and provide insights into the problem solution
- Perform decision-making within the context of systems

Most problems exist in the system context.

**Definition**

A *system* is a set of interrelated components, which make up a united whole. Each component interacts with or is related to at least one other component, and they all serve a common objective.

*Environment* is the aggregate of elements that do not belong to the system but influence it.

*Structure* is the characteristic of steady links and interaction modes of a system's elements; structure determines the system's integrity, organization, and functioning; structure is the framework of a system.

*Dynamics* is a system's change and evolution in object and process states in space and time as driven by the perturbation and control influences resulting from both planned transition from a current state to a desired one and adaptation of a system to a changed execution environment.

*Decision* is the selection of an activity (or a set of activities) from several alternatives. Decisions are subject to constraints, which limit decision choices and objectives which make some decisions preferable to others.

Managerial decisions affect system goal orientation through output performance. Decisions shape system behavior with regard to a certain goal (or multiple goals). Objectives and criteria play the most important role in making decisions. Each management decision has two components. The first one is an *analytical* component, and the second one is a *behavioral* component. The analytical component is supported by *quantitative analysis* business analytics methods, while the behavioral component is based on the intuition and leadership qualities of the decision-maker as well as on external environment *behavior prediction and reaction* in regard to the decisions of suppliers, retailers, and customers.

*Performance* is a complex characteristic of the goal-oriented results of the system's functioning. Managers must consider potential and real performance and take into account the conformity of the achieved results with the goals set at the planning stage.

Performance is measured with certain metrics (e.g., capacity utilization, on-time delivery, and costs), which are frequently called as *key performance indicators* (KPIs).

With the developments in digital technology and artificial intelligence, decisions in SCM are frequently taken in a hybrid mode combining human and artificial intelligence. The decisions become more and more data driven.

### 1.2.2 Models and Data

The concept of a model is widely applicable in natural human languages and is a general scientific term. It is characterized by polysemy, that is, widely expressed and reflecting different meanings depending on the applications and contexts.

**Definition**

A *model* is:

- Simplified presentation of a real object/system
- A system whose investigation is a tool for obtaining information about another system
- A method of knowledge existence
- A multiple system map of the original object that, together with absolutely true content, contains conditionally true and false content, which reveals itself in the process of its creation and practical use

*Modeling* is one of the stages of cognitive activity of a subject, which involves the development (choice) of a model, use of it to conduct investigations, obtainment and analysis of the results, production of recommendations, and estimation of the quality of the model itself as applied to the solved problem and taking into account specific conditions (◘ Fig. 1.5).

Because of the finiteness of the designed (applied) model (a limited number of elements and relations that describe objects belonging to infinitely diverse reality) and the limited resources (temporal, financial, and material) supplied for modeling, the model always reflects the original object in a simplified and approximate way. However, experience shows that these specific features of a model are admissible and do not oppose the solution of problems that are faced by subjects.

Modeling quality depends on *data quality*. Data can be classified according to several perspectives as follows:

- Time: *past data* (historical data, e.g., sales for the last 3 years), *real-time data* (e.g., current inventory level at a warehouse), and *forecasted data* (e.g., sales forecast)

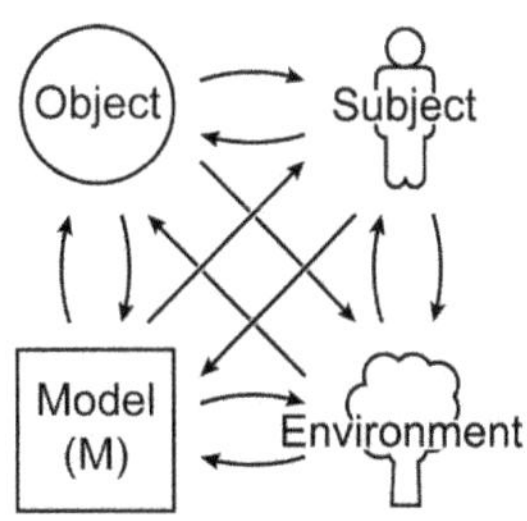


◘ **Fig. 1.5** Modeling stakeholders (Ivanov & Sokolov, 2010)

1

- Origin: *primary data* (e.g., real lead time data from an enterprise resource planning (ERP) system) and *synthetic data* (i.e., data obtained through training in an artificial intelligence tool)
- Content: *qualitative data* (e.g., expert opinions) and *quantitative data* (e.g., production capacity)

### 1.2.3 Data-Driven, Model-Based Decision-Making Support

The model-based decision-making process is shown in ◘ Fig. 1.6.

We can observe that a *real management problem* is the initial point of the decision-making process. An example might be a facility location problem where we are given demand in some markets, possible locations and capacities of new facilities, fixed costs of having a facility in the supply chain, and transportation costs from each location to each market. We must decide where to locate the facilities and which quantities should be shipped from the facilities to the markets.

The next step is to transform the real problem into a *mathematical model*. For this transformation, we need to reduce the *complexity* of reality. This inevitably results in simplification. For example, we would assume deterministic capacity in our facility location model instead of considering fluctuations in demand.

We make this simplification in order to represent the management problem in such a way that the model can be solved with the help of existing *algorithms* within a reasonable time. In our example, we formulate the facility location problem as a mixed-integer linear programming model that can be solved with the help of a branch-and-bound algorithm.

For implementation, *software* is needed. The software calculates a *solution*. For example, a solution could suggest where to open facility locations and which product quantities should be shipped from each opened location to each of the markets so that total production and logistics costs are minimal.

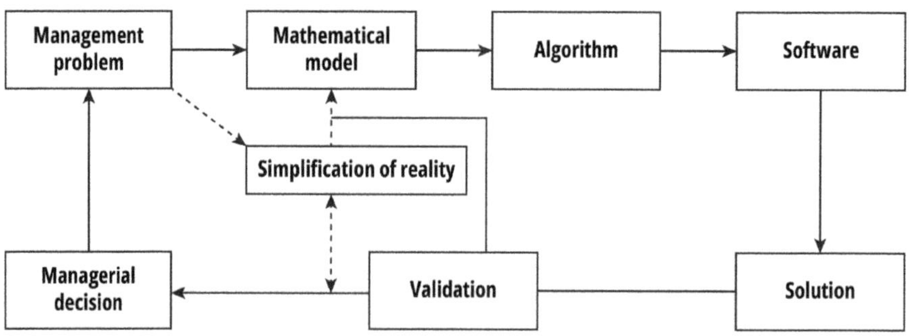


◘ **Fig. 1.6** Model-based decision-making process

**Important Observation**

The software calculates this solution. The most important question now is whether this solution is automatically our *decision*? No! This is a solution to the mathematical problem. First, this solution should be validated by managers (e.g., through a sensitivity analysis and comparison with some performance statistics at the company). If the results are not valid, the model should be corrected, and full modeling cycle should be repeated. Second, management expertise is needed to transfer this mathematical solution into managerial decisions. In this way, data-driven, model-based supply chain analytics work.

## 1.3 Methods of Supply Chain Analytics

Supply chain analytics methods span *operations research* and *artificial intelligence* areas. This section aims to navigate the reader through the variety of the existing methods providing some indicative literature for different methods. Throughout the book, we will explain some of the supply chain analytics methods in more detail. ◘ Figure 1.7 illustrates the variety of supply chain analytics methods structuring them according to the underlying mathematical and statistics methodologies.

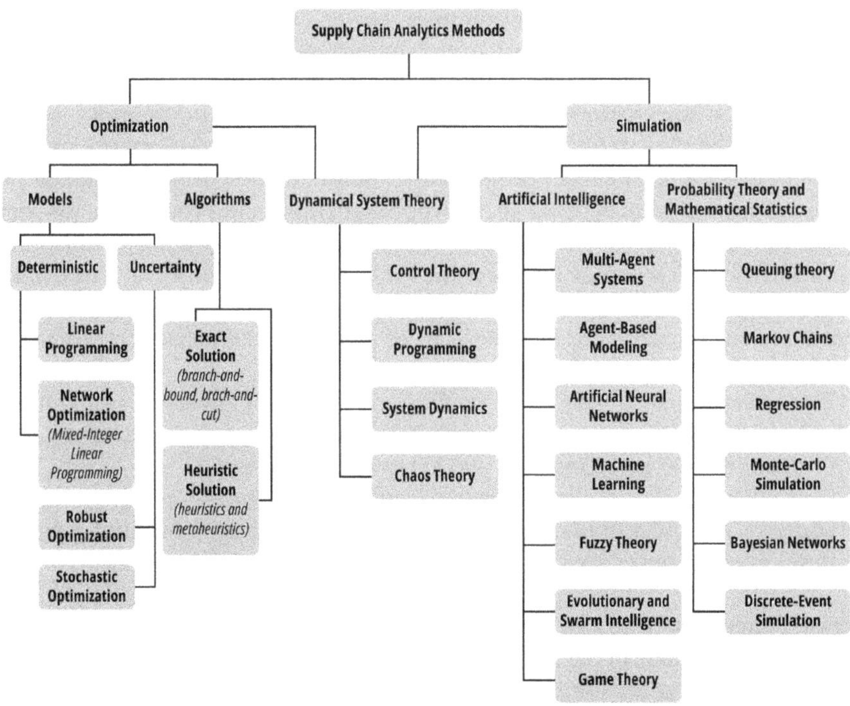


◘ **Fig. 1.7** Classification of supply chain analytics methods

Supply chain analytics are built on the following two major methods:
- *Optimization*
- *Simulation*

*Optimization* methods are used to solve a management problem representing it through a set of equations. *Simulation* methods are based on observing systems' behavior in time based on some rules defined by a modeler. These rules can be derived using probability theory, mathematical statistics, analytical modeling, and artificial intelligence. A special class of supply chain analytics methods is dynamical systems theory, which combines elements of optimization and simulation to define and observe linear and nonlinear system behaviors in time.

## 1.4 Mathematical Optimization

*Mathematical optimization* methods help to find optimal solution to a problem expressed through a model that is composed of an *objective function*, a set of given *parameters* (e.g., costs, demand, and capacities), some *decision variables* (e.g., transportation flows from factories to DCs), and a set of *constraints* (e.g., production capacity constraint). Objective functions can have a single criterion (e.g., profit maximization) or be *multi-criteria* (e.g., both cost minimization and resilience maximization). The models are solved through specific algorithms, which can find the best possible (i.e., optimal) solution (e.g., *branch-and-bound* used in CPLEX solver) or a very good, near-optimal solution (e.g., *metaheuristics*). The results of the model solution are expressed through decision variables. For example, *network optimization* (NO) models find optimal flows in the supply chain and optimal supply chain designs (Simchi-Levi et al., 2007; Ivanov & Dolgui, 2021). *Linear programming* models can be used to find optimal (from the profit point of view) production quantities (Render et al., 2014). Depending on data, mathematical optimization models can be *deterministic* (i.e., all the parameters such as demand and capacity are considered to be known) or *uncertain* (i.e., some parameters such as demand and capacity are considered to be unknown or stochastic) (Mula et al., 2010).

The *optimal decision* is the best one. It is important to know that optimality can be understood in different ways (Ivanov, 2022) subject to several criteria, e.g.:
- Profitability maximization (i.e., maximum effectiveness)
- Cost-efficiency (i.e., minimum resource consumption)
- Usage of resources in the most sustainable way (i.e., maximum sustainability)
- Resilience maximization

Another important aspect of optimality is the constraints of the model. The more constraints we include in the model, the more narrow the solution space potentially excluding a very good solution. As such, only the most important constraints should be considered.

In some cases, *multi-criteria optimization* is performed. A particular feature of optimal decisions is multi-objective decision-making by managers with their own preferences that, in their turn, are always changing. Hence, one important task is to build a general selection function for multi-criteria decision-making. A multi-criteria decision-making strategy employs *trade-off* analysis techniques to identify compromise designs for which the competing criteria are mutually satisfied in a *pareto-optimal* sense. For more detailed information on multi-objective decision-making theory, we refer to the studies on operations research, decision theory, and quantitative analysis in management, e.g., Taha (2009), Render et al. (2014), Dolgui and Proth (2010), and Yalaoui et al. (2012).

## 1.5 Simulation

### 1.5.1 Probability Theory and Mathematical Statistics

*Probability theory* and *mathematical statistics* provide a number of methods which are used in supply chain analytics when we face problems with *randomness* of events and *uncertainty*. These methods help us analyze and interpret data to extract information, build discrete event-driven models, and make predictions. While probability deals with predicting the likelihood of future events, statistics is concerned with the analysis of the past events.

*Queuing theory* helps to analyze a performance of a process for problems with random arrival of tasks in a system (e.g., random arrival of customers to a service center). Queuing theory builds a base for *discrete event simulation*, which is used for process analysis and planning (Ivanov et al., 2021b; Borshchev & Grigoryev, 2023). The principles and tools of discrete event simulation are based on considering some input (e.g., customer arrivals) and transformation (e.g., customer service) events and their linear interrelations, mostly based on queuing theory. Discrete event models are frequently used for inventory control in supply chains (Ivanov, 2020). *Regression analysis* helps determine the impact of one variable on another one (e.g., how much would our sales of ice cream increase in summer when the demand is supposed to increase compared to winter). *Markov chain* is a probabilistic graphical model describing a sequence of possible events. The probability of each event depends only on the state attained in the previous event. One specific case of Markov chains for linear systems is *Bayesian networks*, which have frequently been used in SCM for resilience analysis (e.g., prediction of supply chain disruption probability based on a supplier disruption probability) (Hosseini & Ivanov, 2020). *Monte Carlo simulation* is a technique allowing to predict some possible outcomes of an uncertain event based on past data and a choice of action. For example, we can use Monte Carlo simulation to analyze the impact of demand fluctuations on supply chain on-time delivery performance for some network design.

1

### 1.5.2 Artificial Intelligence

*Artificial intelligence* (AI) is a simulation of human decision-making processes by machines and computers.

Humans started using computers for decision-making support a long time ago (Wiener, 1948). Even during its infancy, AI tried to mimic human decision rules and system behaviors and has been considered an important part of decision-making support. Thanks to rapid technological developments, modern AI approaches such as deep learning, reinforcement learning, federated learning, neural networks, and visual and interactive intelligence seem to overcome the limitations of early-stage AI (Kosasih & Brintrup, 2022; Rolf et al., 2022). They not only mimic human-defined decision rules (which might be inaccurate due to human's limited ability to describe their decision-making rules) but also create new knowledge through training and frequently outperform humans (Fügener et al., 2022). Most importantly, modern technologies allow to develop human-AI collaboration based on mutual delegating tasks and decision-making authorizations between humans and AI, which is referred to as "human-in-the-loop artificial intelligence" (Zanzotto, 2019).

◘ Figure 1.8 provides an overview of AI-based methods.

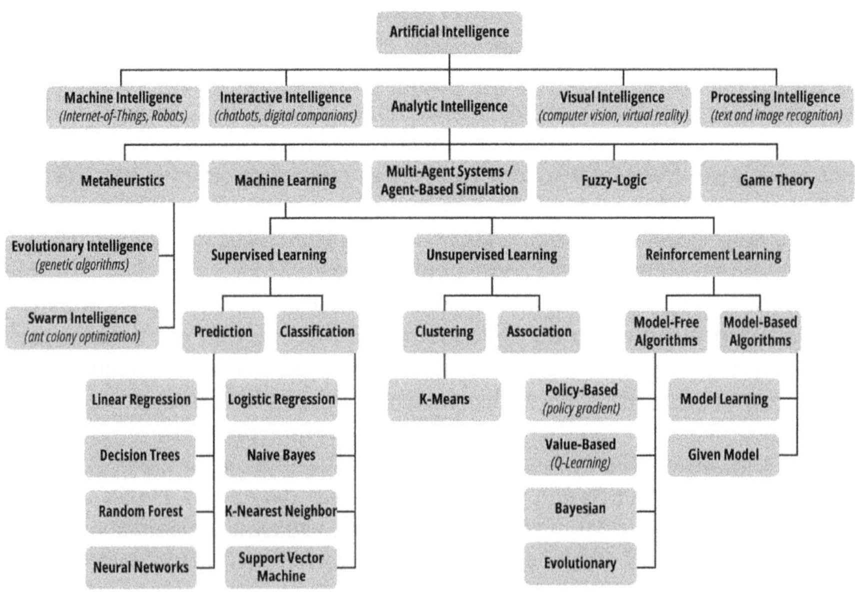


◘ **Fig. 1.8** Overview of AI-based methods

We particularly focus on the analytic intelligence group. *Multiagent systems* and *agent-based modeling* represent a system of intelligent entities (i.e., agents), each of which acts based on some rules, interests, and behaviors interacting with other agents. Agent-based simulation modeling is frequently used in SCM to examine competition and collaboration practices with the use of complex adaptive system principles. For example, logistics service providers can be described as competing agents bidding on transportation orders. The agents can not only compete but also collaborate, e.g., in order to maximize the overall profit (Swaminathan et al., 1998; Surana et al., 2005; Nair & Vidal, 2010; Ivanov, 2018).

*Machine learning* methods are based on training principles (Liu, 2022). Algorithms which imitate human decision-makers are trained and learn through this training. Machine learning algorithms can be supervised (e.g., labeled datasets are used to train algorithms for accurate data classification or outcome predictions) and unsupervised (e.g., analysis and clustering of unlabeled datasets). One specific area of machine learning is *reinforcement learning* that combines Markov chain and agent-based simulation elements. Intelligent agents in reinforcement learning algorithms interact with an environment and learn to act within this environment. Supervised machine and reinforcement learning have been applied to SCM in different areas, e.g., inventory management, transportation, and supplier selection (Rolf et al., 2022). For example, an agent can learn how to place optimal orders or to select suppliers with highest delivery reliability. Machine learning techniques are frequently combined with discrete event simulation (Cavalcantea et al., 2019).

*Artificial neural networks* are inspired by biological neural network framing mechanisms of animal brains. Neural network models are based on machine learning and training. *Fuzzy theory* is used when some of the modeling elements cannot be quantified and should be described verbally, e.g., high, medium, or low demand. *Evolutionary algorithms* (i.e., metaheuristics such as genetic algorithms) are used for solving supply chain optimization problems when exact solutions are impossible due to problem complexity (so-called NP-hard problems) (Mirjalili, 2018). *Game theory* is a branch of mathematics dealing with interactions of agents equipped with some rational and goal-oriented behavior. Game theory is very popular in the analysis of negotiations and contracting in supply chains (Gupta et al., 2021).

Four other groups shown in ◘ Fig. 1.8 (i.e., machine intelligence, interactive intelligence, visual intelligence, and processing intelligence) also find numerous applications in SCM. Through machine intelligence, cobots (collaborative robots) help improve manufacturing processes by mutual enhancements of human and robot intelligence (Sheu & Choi, 2023). Visual intelligence creates new opportunities for companies to interact with customers in digital world, e.g., *metaverse* (Dolgui & Ivanov, 2023). Use of processing intelligence (e.g., natural language processing such as ChatGPT) allows to simplify simulation model development through automatic coding (Jackson et al., 2024).

## 1.6 Dynamical System Theory

*Dynamical systems* are comprised of optimal control, dynamic programming, system dynamics, and chaos theory. The major distinguishing principle of these methods is consideration of time at a highly granular level, i.e., the consideration of system and process dynamics. *Control theory* has been developed to study multi-stage, multi-period dynamic systems over time. Control theory contains a rigor quantitative basis for planning optimal control policies including differential games and stochastic systems, feedback control, stability of controlled processes and nonlinear systems, controllability and observability, and adaptation (Sethi & Thompson, 2000; Disney et al., 2006; Dolgui et al., 2019). Applications of control theory, simulation, and systems science to supply chain dynamics are multifaceted covering inventory management, production scheduling, and supply chain resilience analysis (Ivanov et al., 2018).

Within control theory, different control approaches exist, i.e., optimal program control (i.e., continuous optimization), model predictive control, and discrete control theory. Model predictive control is a control strategy based upon the explicit use of a process model to predict the process output (performance) over a long period of time. The model attempts to predict the control variables for a set of time periods. Predicted control variables depend on disturbance forecasts (i.e., demand, prices, and interest rates) and also on a set of given parameters, so-called control inputs. A system model and current and historical measurements of the process are used to predict system behavior at future time instants. A control-relevant objective function is then optimized to calculate a sequence of future control moves that must satisfy the system constraints.

Feedback-based *discrete control theory* can be efficiently applied to production inventory control problems (Disney & Towill, 2002). The major idea of dynamic production inventory control applications is to present the operational system and closed-loop feedback cycles using a methodology of transfer function and state-space models. The bullwhip effect and stability of supply chains are other typical application areas of the linear control theory (Disney et al., 2008).

*Optimal program control* is devoted to determining some functions known as controls that lead to optimization (minimization or maximization) of an objective (e.g., cost minimization). Optimal program control takes a different perspective as mathematical programming methods representing supply chain flows as dynamic trajectories rather than discrete flows (Dolgui et al., 2019; Ivanov et al., 2021b). Two main computational principles for the optimal program control models are the *Pontryagin's maximum principle* (Ivanov & Sokolov, 2010, Sect. 12.3) and the *dynamic programming* method (Bellmann, 1972). Optimal program control can be efficiently used for optimal design, planning, and scheduling of complex multi-stage, multi-period, and multi-commodity supply chains. However, an important precondition is the existence of centralized information for the whole network and relevant parameters. Therefore, optimal control can be efficiently applied for centralized network structures and can require some extensions with other methods (e.g., agent-based models and machine learning) for decentralized decision-making (Ivanov, 2009; Ivanov et al., 2019).

Nonlinear models based on differential equations are applied as *system dynamics* models to represent the SC, manufacturing, and logistics processes using stocks, flows, internal feedback loops, table functions, and time delays (Sterman, 2000). The objective is to understand the nonlinear behaviors in complex networks. System dynamics models normally relate relatively rough to real problems and very abstractly. In particular, they find applications in supply chain resilience area (Ghadge et al., 2022). *Chaos theory* deals with the behavior of specific dynamical systems whose states evolve over time with a high sensitivity to initial conditions. For example, chaos theory can be used to determine dynamic instabilities in supply chains (Demirel et al., 2019).

## 1.7 Applications of Supply Chain Analytics: Descriptive, Predictive, and Prescriptive

Application areas of supply chain analytics can be classified into three areas, i.e.:
- Descriptive and diagnostic analysis
- Predictive simulation
- Prescriptive optimization

▣ Figure 1.9 illustrates the supply chain analytics areas.

*Descriptive analytics* can help observe retrospective dynamics through performance visualization.

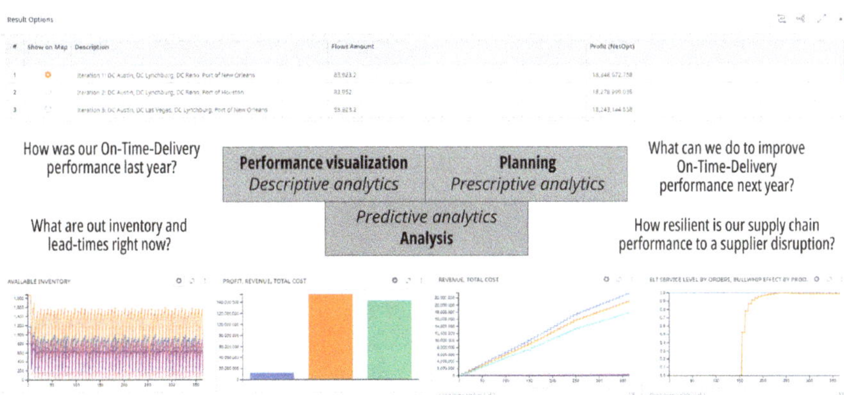


▣ **Fig. 1.9** Supply chain analytics areas

1

► Example

For example, a descriptive analytics method can be used to compute that say 90% of our orders in the past arrived on time, and 10% of our orders were late. Descriptive analytics can also be used for the analysis of the current situation in the SC. Real-time data about inventory levels, capacity utilization, and truck locations obtained in the digital supply chain from, e.g., radio-frequency identification (RFID), blockchain, and product sensors can be used to describe current product availability in the warehouses, production capacity utilizations, and delivery performance (delayed or on time). However, descriptive analytics do not explain why 10% of our orders were late, why our product availability has decreased, and what on-time delivery performance we can expect next year. ◄

To answer these questions, *predictive analytics* can be used which are mainly based on simulation methods. They make it possible to analyze the reasons for some events in the past and predict future developments based on the past or real-time data. For example, a simulation model can help predict that our on-time delivery performance will be 95%. ◘ Figure 1.10 summarizes major network, process, and behavior simulation models used in SCM.

Discrete event simulation models are used for process analysis. Agent-based simulation allows to represent human thinking and behaviors. System dynamics simulation makes it possible to study complex network behaviors. However, the predictive analytics do not tell us what we should do to improve our on-time delivery performance and reach 95%.

At this stage, *prescriptive analytics* need to be considered which help to reveal the course of future actions. Prescriptive analytics are based on optimization methods, some examples of which are shown in ◘ Fig. 1.11.

We refer to ► Sects. 1.4 and 1.6 for detailed description of different methods shown in ◘ Fig. 1.11. For example, an NO model can determine that we should open a new warehouse to increase on-time delivery from 90% to 95%. ◘ Table 1.1 illustrates some practical applications of data analytics to SCM using a simple example of on-time delivery performance indicator (i.e., the percentage of on-time fulfilled orders).

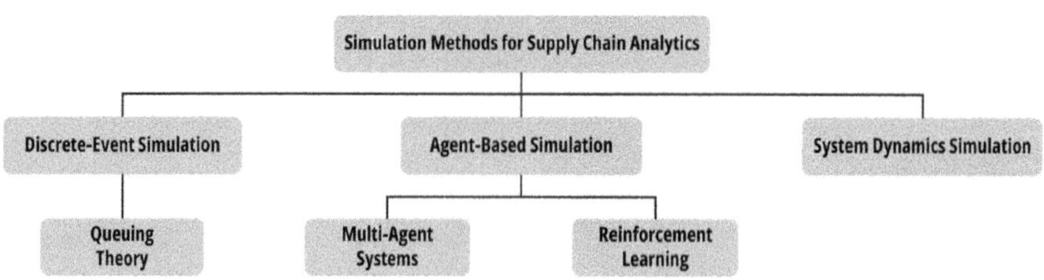


◘ **Fig. 1.10** Major network, process, and behavior simulation models used in SCM

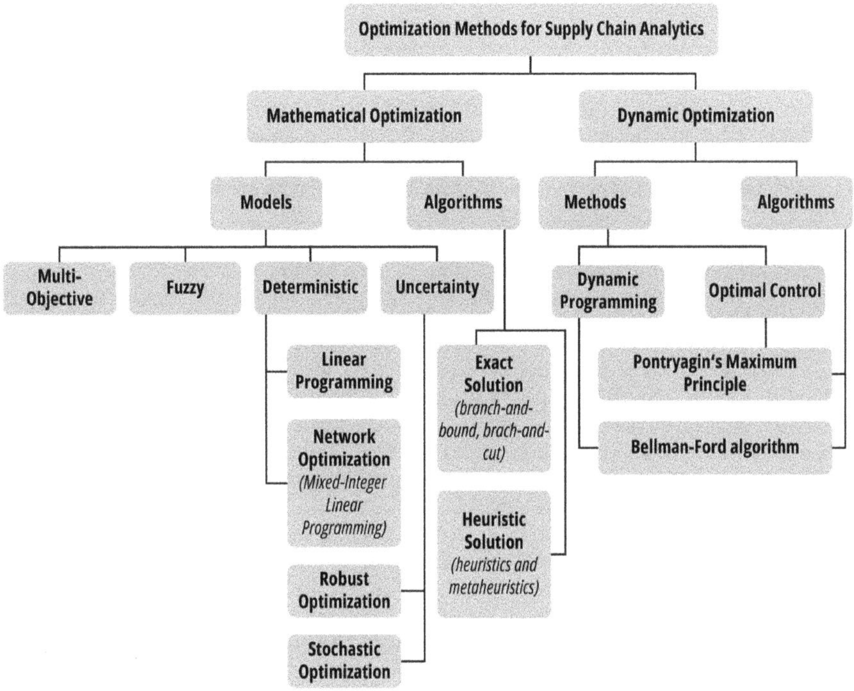


**Fig. 1.11** Examples of optimization methods used in supply chain analytics

**Table 1.1** Examples of data analytics applications to SCM

| | Descriptive analytics | Predictive analytics | Prescriptive analytics |
|---|---|---|---|
| Key question | What happened in the past? | What can happen in future? | What is our course of action? |
| Method | Data visualization | Simulation | Network optimization |
| Analysis outcome | On-time delivery of 90% last year | Next year, we can expect on-time delivery of 95% | We should open a new warehouse to increase on-time delivery from 90% to 95% |

## References

Bellmann, R. (1972). *Adaptive control processes: A guided tour*. Princeton University Press.

Borshchev, A., & Grigoryev, I. (2023). The big book of simulation modeling. Retrieved from https://www.anylogic.com/resources/books/big-book-of-simulation-modeling/

Cavalcantea, I. M., Frazzon, E. M., Forcellinia, F. A., & Ivanov, D. (2019). A supervised machine learning approach to data-driven simulation of resilient supplier selection in digital manufacturing. *International Journal of Information Management, 49*, 86–97.

Daellenbach, H. G., & McNickle, D. C. (2005). *Management science. Decision making through systems thinking*. Palgrave Macmillan.

Demirel, G., MacCarthy, B. L., Ritterskamp, D., Champneys, A., & Gross, T. (2019). Identifying dynamical instabilities in supply networks using generalized modeling. *Journal of Operations Management, 65*(2), 136–159.

Disney, S. M., & Towill, D. R. (2002). A discrete linear control theory model to determine the dynamic stability of vendor managed inventory supply chains. *International Journal of Production Research, 40*(1), 179–204.

Disney, S. M., Towill, D. R., & Warburton, R. D. H. (2006). On the equivalence of control theoretic, differential, and difference equation approaches to modeling supply chains. *International Journal of Production Economics, 101*, 194–208.

Disney, S. M., Lambrecht, M., Towill, D. R., & Van de Velde, W. (2008). The value of coordination in a two-echelon supply chain. *IIE Transactions, 40*(3), 341–355.

Dolgui, A., & Ivanov, D. (2023). Metaverse supply chain and operations management. *International Journal of Production Research, 61*(23), 8179–8191. https://doi.org/10.1080/00207543.2023.2240900.

Dolgui, A., & Proth, J.-M. (2010). *Supply chain engineering: Useful methods and techniques*. Springer.

Dolgui, A., Ivanov, D., Sethi, S. P., & Sokolov, B. (2019). Scheduling in production, supply chain and Industry 4.0 systems by optimal control. *International Journal of Production Research, 57*(2), 411–432.

Fügener, A., Grahl, J., Gupta, A., & Ketter, W. (2022). Cognitive challenges in human–artificial intelligence collaboration: Investigating the path toward productive delegation. *Information Systems Research, 33*(2), 678–696.

Ghadge, A., Er, M., Ivanov, D., & Chaudhuri, A. (2022). Visualisation of ripple effect in supply chains under long-term, simultaneous disruptions: A system dynamics approach. *International Journal of Production Research, 60*(20), 6173–6186.

Gupta, V., Ivanov, D., & Choi, T.-M. (2021). Competitive pricing of substitute products under supply disruption. *Omega, 101*, 102279.

Hosseini, S., & Ivanov, D. (2020). Bayesian networks for supply chain risk, resilience and ripple effect analysis: A literature review. *Expert Systems with Applications, 161*, 113649.

Ivanov, D. (2009). DIMA—A research methodology for comprehensive multi-disciplinary modelling of production and logistics networks. *International Journal of Production Research, 47*(5), 1153–1173.

Ivanov, D. (2018). *Structural dynamics and resilience in supply chain risk management*. Springer.

Ivanov, D. (2020). Predicting the impacts of epidemic outbreaks on global supply chains: A simulation-based analysis on the coronavirus outbreak (COVID-19/SARS-CoV-2) case. *Transportation Research Part E: Logistics and Transportation Review, 136*, 101922.

Ivanov, D. (2022). Lean resilience: AURA (active usage of resilience assets) framework for post-COVID-19 supply chain management. *International Journal of Logistics Management, 33*(4), 1196–1217.

Ivanov, D., & Dolgui, A. (2021). OR-methods for coping with the ripple effect in supply chains during COVID-19 pandemic: Managerial insights and research implications. *International Journal of Production Economics, 232*, 107921.

Ivanov, D., & Sokolov, B. (2010). *Adaptive supply chain management*. Springer.

Ivanov, D., Sethi, S., Dolgui, A., & Sokolov, B. (2018). A survey on the control theory applications to operational systems, supply chain management and Industry 4.0. *Annual Reviews in Control, 46*, 134–147.

Ivanov, D., Dolgui, A., Ivanova, M., & Sokolov, B. (2019). Intellectualization of control: Cyber-physical supply chain risk analytics. In D. Ivanov, A. Dolgui, & F. Yalaoui (Eds.), *9th IFAC Conference on Manufacturing Modelling, Management and Control (MIM 2019)* (Vol. 52(13), pp. 355–360). IFAC-PapersOnLine.

Ivanov, D., Tsipoulanidis, A., & Schönberger, J. (2021a). *Global supply chain and operations management: A decision-oriented introduction into the creation of value*. Springer Nature.

Ivanov, D., Sokolov, B., Chen, W., Dolgui, A., Werner, F., & Potryasaev, S. (2021b). A control approach to scheduling flexibly configurable jobs with dynamic structural-logical constraints. *IISE Transactions, 53*(1), 21–38.

Jackson, I., Saenz, M., & Ivanov, D. (2024). From natural language to simulations: Applying AI to automate simulation modelling of logistics systems. *International Journal of Production Research, 62*(4), 1434–1457.

Kosasih, E. E., & Brintrup, A. (2022). A machine learning approach for predicting hidden links in supply chain with graph neural networks. *International Journal of Production Research, 60*(17), 5380–5393.

Liu, K. Y. (2022). *Supply chain analytics. Concepts, techniques and applications*. Springer Nature.

Mirjalili, S. (2018). *Evolutionary algorithms and neural networks: Theory and applications*. Springer Nature.

Mula, J., Peidro, D., Díaz-Madroñero, M., & Vicens, E. (2010). Mathematical programming models for supply chain production and transport planning. *European Journal of Operational Research, 204*(3), 377–390.

Nair, A., & Vidal, J. M. (2010). Supply network topology and robustness against disruptions—An investigation using a multi-agent model. *International Journal of Production Research, 49*(5), 1391–1404.

Render, B., Stair, R. M., & Hanna, M. E. (2014). *Quantitative analysis for management* (12th ed.). Pearson.

Rolf, B., Jackson, I., Müller, M., Lang, S., Reggelin, T., & Ivanov, D. (2023). A review on reinforcement learning algorithms and applications in supply chain management. *International Journal of Production Research, 61*(20), 7151–7179. https://doi.org/10.1080/00207543.2022.2140221.

Sethi, S. P., & Thompson, G. L. (2000). *Optimal control theory: Applications to management science and economics* (2nd ed.). Springer.

Sheu, J. B., & Choi, T.-M. (2023). Can we work more safely and healthily with robot partners? A human-friendly robot-human coordinated order fulfillment scheme. *Production and Operations Management. 32*(3), 94–812, https://doi.org/10.1111/poms.13899.

Simchi-Levi, D., Kaminsky, P., & Simchi-Levi, E. (2007). *Designing and managing the supply chain: Concepts, strategies and case studies* (3rd ed.). McGraw-Hill.

Sterman, J. (2000). *Business dynamics: Systems thinking and modeling for a complex world*. McGraw-Hill.

Surana, A., Kumara, S., Greaves, M., & Raghavan, U. N. (2005). Supply-chain networks: A complex adaptive systems perspective. *International Journal of Production Research, 43*(20), 4235–4265.

Swaminathan, J. M., Smith, S. F., & Sadeh, N. M. (1998). Modeling supply chain dynamics: A multiagent approach. *Decision Sciences, 29*(3), 607–632.

Taha, H. A. (2009). *Operations research: An introduction* (9th ed.). Pearson.

Wiener, N. (1948). *Cybernetics or control and communication in the animal and the machine*. John Wiley & Sons Inc.

Yalaoui, A., Chehade, H., Yalaoui, F., & Amodeo, L. (2012). *Optimization of logistics*. Wiley.

Zanzotto, F. M. (2019). Human-in-the-loop artificial intelligence. *Journal of Artificial Intelligence Research, 64*, 243–252.

1

## Further Reading


Bertrand, J. W. M., & Fransoo, J. C. (2002). Modelling and simulation: Operations management research methodologies using quantitative modeling. *International Journal of Operations & Production Management, 22*(2), 241–264.

Bozarth, C. B., & Handfield, R. B. (2019). *Introduction to operations and supply chain management* (5th ed.). Pearson.

Casti, J. L. (1979). *Connectivity, complexity and catastrophe in large-scale systems*. Wiley-Interscience.

Choi, T. Y., Dooley, K. J., & Rungtusanatham, M. (2001). Supply networks and complex adaptive systems: Control versus emergence. *Journal of Operations Management, 19*, 351–366.

Choi, T. M., Wallace, S. W., & Wang, Y. (2018). Big data analytics in operations management. *Production and Operations Management, 27*(10), 1868–1883.

Chopra, S., & Meindl, P. (2015). *Supply chain management: Strategy, planning and operation* (6th ed.). Pearson.

Lee, E. B., & Markus, L. (1967). *Foundations of optimal control theory*. John Wiley & Sons.

Melo, M. T., Nickel, S., & Saldanha-da-Gama, F. (2009). Facility location and supply chain management—A review. *European Journal of Operational Research, 196*(2), 401–412.

Mesarovic, M. D., & Takahara, Y. (1975). *General systems theory: Mathematical foundations*. Academic Press.

Oliver, R. K., & Webber, M. D. (1982). Supply chain management: Logistics catches up with strategy. In M. Christopher (Ed.), *Logistics: The strategic issues*. Chapman & Hall.

Rozhkov, M., Ivanov, D., Blackhurst, J., & Nair, A. (2022). Adapting supply chain operations in anticipation of and during the COVID-19 pandemic. *Omega, 110*, 102635.

# Demand Forecasting, Production Planning, and Inventory Control

Contents




D. Ivanov, *Introduction to Supply Chain Analytics*, Classroom Companion: Business,
https://doi.org/10.1007/978-3-031-51241-4_2

**Learning Objectives**

In this chapter, we introduce major terminology, methods, and applications of supply chain analytics for demand forecasting, production planning, and inventory control. First, we learn how to manage demand and lead time uncertainty. Statistical demand forecasting methods and inventory control policies are explained. Subsequently, production planning using linear programming method is introduced.

To this end, our learning objectives for this chapter are as follows:

- Principles and methods of statistical demand forecasting.
- Inventory control policies.
- Production planning using linear programming.

## 2.1 Forecasting Methods and Accuracy

### 2.1.1 Quantitative Forecasting Methods

Quantitative methods are used when historical data exists, e.g., existing products, existing technology or existing markets. These methods are based on statistics. Statistical methods can be classified in two different *forecasting* methods:

- causal forecast and
- time series analysis.

In *causal forecasting*, the demand ($y$) for a product depends on a known factor ($x$), and because of this, it can be forecast (see ◘ Fig. 2.1). For example, demand for axis ($y$) depends on the number of citizens in a district ($x$). Linear trends can be found using the least squares technique that results in Eq. (2.1):

$$\hat{y} = a + bx \tag{2.1}$$

where $\hat{y}$ is the computed value of the variable to be predicted (dependent variable); $a$ is the $y$-axis intercept; $b$ is the slope of the regression line; and $x$ is the independent variable.

Nonlinear regression is more complicated, but also more realistic. For a causal forecasting method, it is important to have reliable historical data for the known factor (the independent variable).

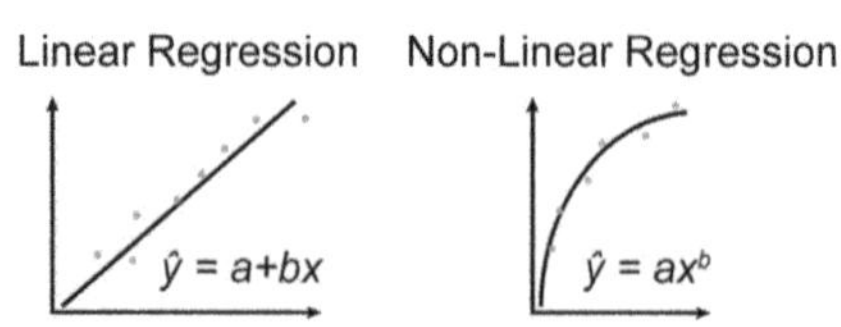


◘ **Fig. 2.1** Regression analysis

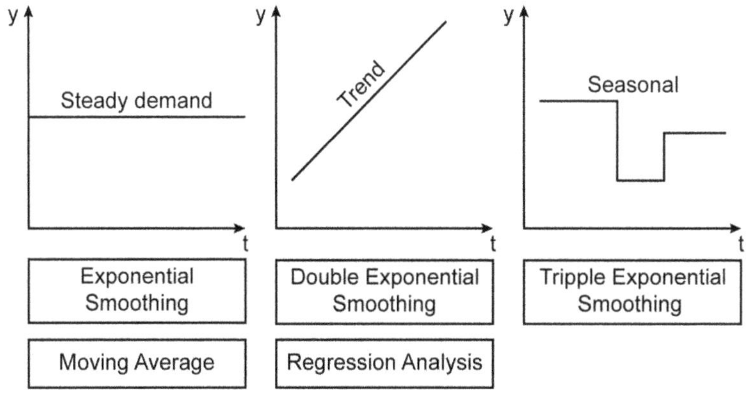


**Fig. 2.2** Time series analysis (Ivanov et al., 2021)

On the other hand, there is *time series analysis*. It is based on historical sales data. Sets of evenly spaced numerical data can be obtained by observing response variables at regular time periods. This method actually forecasts on the basis of the past values and assumes that factors influencing past and present will continue to influence in future.

There are different methods to forecast demand using time series analysis (see Fig. 2.2):

If there is a constant demand for a product, we can use the *moving average* or the *simple exponential smoothing*. For products with a demand which is based on trends, we can use *regression analysis* or *double exponential smoothing*. Products with seasonal demand can be forecast using *triple exponential smoothing*.

### 2.1.2 Forecasting Accuracy Quality

Forecasting methods can provide a different *forecast quality*. In order to estimate the quality of a forecast, some measures are used in practice, including:

- Mean absolute deviation (MAD)
- Mean squared error (MSE)
- Mean absolute percentage error (MAPE)

MAD measures the absolute deviation ($\varepsilon$) of forecast and actual data over $T$-periods according to formula (2.2):

$$\text{MAD} = \frac{1}{T}\sum_{t=1}^{T}|\varepsilon_t| \tag{2.2}$$

MSE measures the quadratic deviation of forecast and actual data according to formula (2.3):

$$\text{MSE} = \frac{1}{T}\sum_{t=1}^{T}\varepsilon_t^2 \tag{2.3}$$

MAPE measures the percentile deviation of forecast and actual data according to formula (2.4):

2

$$\mathrm{MAPE} = \frac{1}{T}\sum_{t=1}^{T}\left|\frac{\varepsilon_t}{y_t}\right| \tag{2.4}$$

The practical problem with the MAD measure is that it expresses the absolute values and is therefore quantity dependent. This makes it difficult to apply MAD for comparison of forecasts for different products. MSE is frequently used because of its good theoretical properties but may be negatively affected by so-called outliers or freak values. MAPE is intuitively good and understandable since it shows the forecasting error as percentage and can be used for comparison of different forecasts where MAD/MSE values differ.

In practice, statistically computed measures for forecast quality analysis need to be checked carefully. For example, consider actual sales data for five periods: 10–10–11–10–11 and two forecasts: (a) 8–8–9–8–9 and (b) 8–12–9–12–9. From a statistical point of view, both forecasts have the same MAD = 2. From the management point of view, the first forecast (a) has a bias since we always produce less than demand. For managers, it is more important to identify bias rather than striving for minimal forecast error.

### 2.1.3 Linear Regression

*Linear regression* is normally used as a causal forecasting method. The procedure will be explained by taking sales in a newly established, small tea shop that depends on the number of customers as an example, as shown in ◘ Table 2.1.

The first step is to calculate the regression function $\hat{y}(x) = \hat{a} + \hat{b}x$ with the parameters $\hat{a}$ and $\hat{b}$. It is important to find the $\hat{a}$ and $\hat{b}$ where the MSE has its minimum. To find the parameters we need to fill out ◘ Table 2.2.

◘ **Table 2.1** Initial data for linear regression

| Week $n$ | Customer $x_n$ | Sales (in €) $y_n$ |
|---|---|---|
| 1 | 26 | 269 |
| 2 | 35 | 389 |
| 3 | 43 | 432 |
| 4 | 67 | 536 |
| 5 | 78 | 685 |
| 6 | 81 | 709 |

**Table 2.2** Results for calculating parameters

| Week $n$ | Customer $x_n$ | Sales $y_n$ (in €) | $x_n^2$ | $x_n \times y_n$ |
|---|---|---|---|---|
| 1 | 26 | 269 | 676 | 6994 |
| 2 | 35 | 389 | 1225 | 13,615 |
| 3 | 43 | 432 | 1849 | 18,576 |
| 4 | 67 | 536 | 4489 | 35,912 |
| 5 | 78 | 685 | 6084 | 53,430 |
| 6 | 81 | 709 | 6561 | 57,429 |
| Σ | 330 | 3020 | 20,884 | 185,956 |

In order to calculate the coefficient $\hat{b}$, formula (2.5) can be used:

$$\hat{b} = \frac{(6 \times 185{,}956) - (330 \times 3020)}{(6 \times 20{,}884) - (330)^2} = \frac{119{,}136}{16{,}404} \approx 7.263 \tag{2.5}$$

The computed value 7.263 means that when there is one more customer per week, sales income rises by approximately 7.26€.

There are two options to calculate $\hat{a}$. The first option is the usage of Eq. (2.6):

$$\hat{a} = \frac{\sum_{n=1}^{N} x_n^2 \sum_{n=1}^{N} y_n - \sum_{n=1}^{N} x_n \sum_{n=1}^{N} (x_n \times y_n)}{N \sum_{n=1}^{N} x_n^2 - \left(\sum_{n=1}^{N} x_n\right)^2} \tag{2.6}$$

The easier way is as follows (Eq. 2.7):

$$\hat{a} = \overline{y} - \hat{b}\overline{x} \tag{2.7}$$

When using formula (2.7), we have to calculate the average values: $\overline{y} = \frac{\sum_{n=1}^{N} y_n}{N}$ and $\overline{x} = \frac{\sum_{n=1}^{N} x_n}{N}$. For our example:

$$\hat{a} = \frac{3020}{6} - \left(7263 \times \frac{330}{6}\right) \approx 103.868$$

The regression function now reads as follows: $\hat{y}(x) = 103.868 + 7.263x$.

The next step is to calculate the MSE using the regression function (see Table 2.3).

To calculate the forecast error $\varepsilon_n$, we use formula (2.8):

$$\varepsilon_n = y(x_n) - y_n \tag{2.8}$$

2

Table 2.3 Results for calculating MSE

| Week $n$ | Customer $x_n$ | Sales $y_n$ (in €) | $y(x_n)$ | $\varepsilon_n$ | $\varepsilon_n^2$ |
|---|---|---|---|---|---|
| 1 | 26 | 269 | 292.706 | 23.706 | 561.974 |
| 2 | 35 | 389 | 358.073 | −30.927 | 956.479 |
| 3 | 43 | 432 | 416.177 | −15.823 | 250.367 |
| 4 | 67 | 536 | 590.489 | 54.489 | 2969.051 |
| 5 | 78 | 685 | 670.382 | −14.618 | 213.686 |
| 6 | 81 | 709 | 692.171 | −16.829 | 283.215 |
| Σ | 330 | 3020 | | | 5234.773 |

To calculate the MSE, we use formula (2.3):

$$\text{MSE} = \frac{5234.773}{6} = 872.462$$

## 2.1.4 Moving Average

The method of *moving average* is based on computing the average demand from the previous periods and using these values for forecasting. It is called "moving" because with every period the average value changes or moves onward.

Consider the following example and function to understand forecasting with the help of the moving average. Here, too, it is important to find the result with the lowest MSE. The demand for each subsequent period is computed subject to formula (2.9):

$$\hat{y}_{t+1} = \frac{1}{T} \sum_{\tau=t-T+1}^{t} y_t \tag{2.9}$$

where $\hat{y}_{t+1}$ is the forecast for the next period, $y_\tau$ is the demand in period $t$, and $T$ is the number of periods. It is important to choose a $T$ that is not too large and not too small. Normally, you calculate with different $T$-values to find the one with the lowest MSE.

Consider this example. Company $X$ wants to find its weekly average demand for cupcakes. Table 2.4 illustrates the last weeks' demand.

Three different numbers of periods to be averaged are given:

$T_1 = 2$; $T_2 = 3$; and $T_3 = 4$. The calculation results are presented in Table 2.5.

**Table 2.4** Initial data for moving average

| Week $t$ | 1 | 2 | 3 | 4 | 5 | 6 | 7 | 8 |
|---|---|---|---|---|---|---|---|---|
| Demand $y_t$ | 26 | 35 | 28 | 42 | 38 | 45 | 46 | 39 |

**Table 2.5** Results calculating moving average with three different $T$s

| $T$ | $y_t$ | $\hat{y}(T_1 = 2)$ | $\hat{y}(T_2 = 3)$ | $\hat{y}(T_3 = 4)$ |
|---|---|---|---|---|
| 1 | 26 | – | – | – |
| 2 | 35 | – | – | – |
| 3 | 28 | 31[a] | – | – |
| 4 | 42 | 32 | 30 | – |
| 5 | 38 | 35 | 35 | 33 |
| 6 | 45 | 40 | 36 | 36 |
| 7 | 46 | 42 | 42 | 39 |
| 8 | 39 | 46 | 43 | 43 |

[a] Forecasted demand from this period onwards

For example, to compute the forecast for week #3: 31 = (26 + 35)/2 → we add demand from the previous two periods and divide it using $T_1$.

To check which $T$ is the one with the smallest MSE, we have to calculate the forecast error. It is important to start with a period where for all $T$-values, the sales have been forecast in order to make the MSE values comparable. In this case, it is week #5 (see Table 2.6).

Using the error calculation formula $\varepsilon_t = \hat{y}(T) - y_t$ and Eq. (2.3), we get

$$\text{MSE}(T_1) = \frac{99}{4} = 24.75 \quad \text{MSE}(T_2) = \frac{122}{4} = 30.5 \quad \text{MSE}(T_3) = \frac{171}{4} = 42.75$$

As we can see, the smallest $T$ is the best one because there the MSE is at its minimum. For this reason, the recommendation for forecasting the next periods will be using $T = 2$.

In some cases, a weighted moving average can be used. In this case, older data usually becomes less important, and the weights are assigned to different periods based on experience and intuition.

**Table 2.6** Results calculating MSE

| $t$ | $y_t$ | $\hat{y}(T_1=2)$ | $\varepsilon_t$ | $\varepsilon_t^2$ | $\hat{y}(T_2=3)$ | $\varepsilon_t$ | $\varepsilon_t^2$ | $\hat{y}(T_3=4)$ | $\varepsilon_t$ | $\varepsilon_t^2$ |
|---|---|---|---|---|---|---|---|---|---|---|
| 1 | 26 | – | – | – | – | – | – | – | – | – |
| 2 | 35 | – | – | – | – | – | – | – | – | – |
| 3 | 28 | 31 | – | – | – | – | – | – | – | – |
| 4 | 42 | 32 | – | – | 30 | – | – | – | – | – |
| 5 | 38 | 35 | −3 | 9 | 35 | −3 | 9 | 33 | −5 | 25 |
| 6 | 45 | 40 | −5 | 25 | 36 | −9 | 81 | 36 | −9 | 81 |
| 7 | 46 | 42 | −4 | 16 | 42 | −4 | 16 | 39 | −7 | 49 |
| 8 | 39 | 46 | 7 | 49 | 43 | 4 | 16 | 43 | 4 | 16 |
| Σ | | | | 99 | | | 122 | | | 171 |

## 2.1.5 Simple Exponential Smoothing

This method is close to that of the weighted moving average. The weightings decline exponentially, and the most recent data are weighted higher. For sales forecasts, average demand/sales values of previous periods are exponentially weighted to forecast the next periods. The simple exponential average means working with a smoothing parameter $\alpha$. The idea is to weight current demand higher and to assign lower weights to the previous demand.

In this method, we use formula (2.10) to forecast the demand for the following periods:

$$\hat{y}_{t,t+1} = \alpha \times y_t + (1-\alpha) \times \hat{y}_{t-1,t} \tag{2.10}$$

where $\hat{y}_{t,t+1}$ is the forecast that we are making in the current period for the following period, $y_t$ is the current demand, and $\hat{y}_{t-1,t}$ is the forecast we made in the last period for the current period.

Consider the following example as an aid to understanding simple exponential smoothing. Demand for a product during the last six periods is provided (Table 2.7).

The smoothing parameters are $\alpha_1 = 0.3$ and $\alpha_2 = 0.5$. The idea is to choose the smoothing parameter which leads to a smaller MSE (see Table 2.8).

For example, to compute the forecast sales for period #3: $54.5 = 0.3 \times 44 + (1 - 0.3) \times 59 \rightarrow$, we multiply the demand in week #2 by $\alpha$ and the forecast demand for week #2 by $(1 - \alpha)$ to get the forecast for week #3. In this case, $\text{MSE}(\alpha_1) = \frac{338.75}{6} \approx 56.46$.

Consider the calculations for Table 2.9:

**Table 2.7** Initial data for single exponential smoothing

| Week $t$ | 1 | 2 | 3 | 4 | 5 | 6 |
|---|---|---|---|---|---|---|
| Demand $y_t$ | 59 | 44 | 61 | 58 | 49 | 52 |

**Table 2.8** Results for single exponential smoothing with $\alpha_1 = 0.3$

| $t$ | $y_t$ | $\hat{y}_{t,t+1}$ | $\varepsilon_t$ | $\varepsilon_t^2$ |
|---|---|---|---|---|
| 0 | | 59 | – | – |
| 1 | 59 | 59 | 0 | 0 |
| 2 | 44 | 59 | 15 | 225 |
| 3 | 61 | 54.5[a] | −6.5 | 42.25 |
| 4 | 58 | 56.45 | −1.55 | 2.40 |
| 5 | 49 | 56.92 | 7.92 | 62.65 |
| 6 | 52 | 54.54 | 2.54 | 6.45 |
| Σ | | | | 338.75 |

[a] Forecasted demand from this period onwards

**Table 2.9** Results for single exponential smoothing with $\alpha_2 = 0.5$

| $t$ | $y_t$ | $\hat{y}_{t,t+1}$ | $\varepsilon_t$ | $\varepsilon_t^2$ |
|---|---|---|---|---|
| 0 | | 59 | – | – |
| 1 | 59 | 59 | 0 | 0 |
| 2 | 44 | 59 | 15 | 225 |
| 3 | 61 | 51.5[a] | −9.5 | 90.25 |
| 4 | 58 | 56.3 | −1.75 | 3.06 |
| 5 | 49 | 57.1 | 8.13 | 66.02 |
| 6 | 52 | 53.1 | 1.06 | 1.13 |
| Σ | | | | 385.46 |

[a] Forecasted demand from this period onwards

2

$\text{MSE}(\alpha_2) = 385.46/6 \approx 64.24$. As we can see, the second smoothing parameter leads to a higher MSE. For this reason, it is advisable in this case to forecast demand for the following periods with the smoothing parameter of 0.3 instead of 0.5.

This method helps to forecast the call for products with a steady demand. But to forecast the sales for products which have a trend-shaped or seasonal demand, we need other methods.

### 2.1.6 Double Exponential Smoothing

Double exponential smoothing is used for situations where a trend already exists. This method works with two smoothing factors, $\alpha$ and $\beta$. The basic idea is that demand is normally overrated when a trend is positive and underrated when the trend is negative. Consider formula (2.11):

$$\hat{y}_{t,t+\tau} = a_t + b_t \times \tau \tag{2.11}$$

where $\hat{y}_{t,t+\tau}$ is the forecast for the following period performed in the current period, $a_t$ is the smoothed forecast, $b_t$ is the smoothed trend parameter (slope), and $\tau$ is the parameter which describes the number of periods for which we are making the forecast. We calculate $a_t$ and $b_t$ as follows from Eqs. (2.12) and (2.13):

$$a_t = \alpha \times y_t + (1-\alpha) c \times (a_{t-1} + b_{t-1}) \tag{2.12}$$

$$b_t = \beta \times (a_t - a_{t-1}) + (1-\beta) \times b_{t-1} \tag{2.13}$$

Assume that we have the demand data of gluten-free bread in a small grocery shop for the last seven periods (◘ Table 2.10).

To forecast the demand for 7 months in the following year, we use the following parameters: $\alpha = 0.2$; $\beta = 0.35$; $\tau = 1$; and $b_1 = 4$. The computational results are presented in ◘ Table 2.11.

By changing the smoothing factors, we can obtain a more reliable forecast for the next periods where a trend exists.

To handle forecasts for seasonal products, we can use *triple exponential smoothing*. In this case, the third parameter, for seasonal fluctuations, is added to double exponential smoothing.

◘ **Table 2.10** Initial data for double exponential smoothing

| Months $t$ | 1 | 2 | 3 | 4 | 5 | 6 | 7 |
|---|---|---|---|---|---|---|---|
| Demand $y_t$ | 249 | 257 | 269 | 285 | 298 | 302 | 312 |

**Table 2.11** Results for double exponential smoothing

| $t$ | $y_t$ | $a_t$ | $b_t$ | $\hat{y}_{t,t+1}$ |
|---|---|---|---|---|
| 1 | 249 | 249 | 4 | |
| 2 | 257 | 253.80 | 4.28 | 253 |
| 3 | 269 | 260.26 | 5.04 | 258.08 |
| 4 | 285 | 269.25 | 6.42 | 265.31 |
| 5 | 298 | 280.14 | 7.99 | 275.67 |
| 6 | 302 | 290.90 | 8.96 | 288.12 |
| 7 | 312 | 302.28 | 9.81 | 299.85 |

## 2.2 Inventory Control

Inventory exists in supply chains because of two main reasons:
- Replenishment in batches—this is *cycle inventory*
- Hedging against uncertainty—this is *safety inventory*

*Cycle inventory* exists as a result of producing or purchasing in *lots* or *batches*. A *lot* or *batch* is the quantity that a firm either produces or purchases at a time. Firms can exploit economy of scale and order in large lots to reduce fixed costs. However, an increase in lot size is usually followed by an increase in inventory carrying costs.

*Safety inventory* is carried to satisfy demand subject to unpredictable demand and lead time fluctuations to reduce product shortages. Safety inventory can help improve product availability in the presence of uncertainty. Safety inventory is concerned with shortage and overage costs. The calculation of safety inventory is based on predetermined *service levels*. Choosing safety inventory involves making a *trade-off* between the costs of having too much inventory and the costs of losing sales due to inventory shortage.

### 2.2.1 Order Quantity Planning for Cycle Inventory

Let us start with the economic order quantity (*EOQ) model*. It is simple and helps us to understand the relationship between ordering and holding costs. Consider a system that exhibits the following characteristics (see Fig. 2.3):
- Demand and lead time are known and constant
- Receipt of inventory is instantaneous and complete
- Quantity discounts are not possible
- The only variable costs are setup and holding
- Stock-out can be avoided

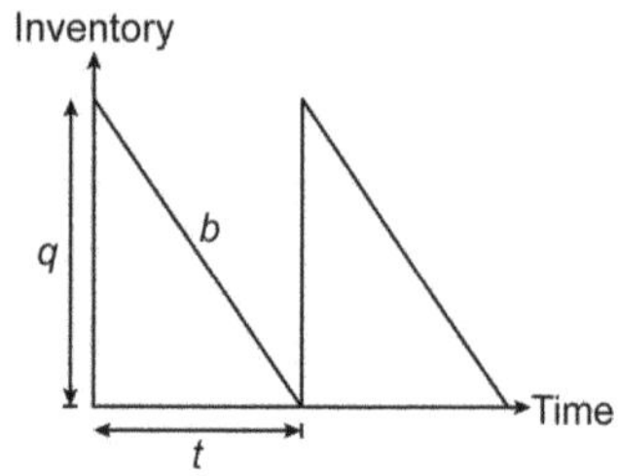


**Fig. 2.3** Inventory consumption and replenishment pattern in EOQ model

We introduce the following notations:
- $q$ is the number of units per order
- $q^*$ is the optimal number of units per order (EOQ)
- $b$ is the annual demand in units for the inventory item
- $f$ is the setup or ordering cost for each order
- $c$ is the holding or carrying cost per unit per year

Under the assumption of linear inventory consumption, cycle inventory and lot sizes are related as follows in Eq. (2.14):

$$\text{Cycle inventory} = \frac{q}{2} \tag{2.14}$$

This means that average inventory on hand is equal to 50% of the order quantity. Then, annual holding cost is $c \times \frac{q}{2}$. In order to calculate ordering costs, we have to know the number of orders per year. This number can be easily calculated as shown in Eq. (2.15):

$$\text{Number of orders per year} = \frac{b}{q} \tag{2.15}$$

$$\text{Then, annual fixed ordering costs is} = f \times \frac{b}{q}$$

Optimal order quantity is found when annual ordering costs equal annual holding costs (see Eq. 2.16):

$$c \times \frac{q}{2} = f \times \frac{b}{q} \tag{2.16}$$

In solving Eq. (2.16) for $q^*$, we get the EOQ formula as follows (Eq. 2.17):

$$q^* = \sqrt{\frac{2b \times f}{c}} \tag{2.17}$$

Consider graphical representation (see Fig. 2.4).

It can be observed from the graph that the total cost (the top curve) is the sum of the two curves below it. Minimal total costs are achieved at the intersection point of the fixed and variable cost curves. This corresponds to the EOQ point $q^*$. It can also be observed that total cost function is quite flat in the minimal region.

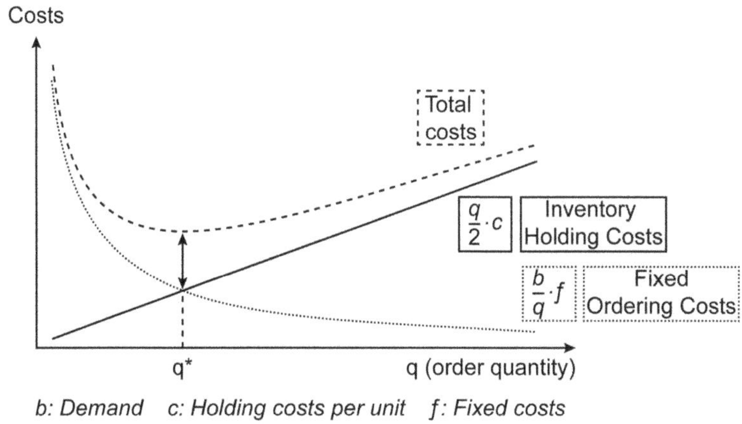


**Fig. 2.4** Graphical representation of EOQ model (Ivanov et al., 2021)

This means that moderate EOQ changes will not influence a significant increase in total costs by tendency. This exemplifies the useful *robustness property* of this method.

We can also determine the expected number of orders per year (Eq. 2.18) and expected time between orders (Eq. 2.19):

$$N = \frac{b}{q^*} \tag{2.18}$$

$$T = \frac{365}{N} \tag{2.19}$$

Total annual cost is calculated as shown in Eq. (2.20):

$$TC = c \times \frac{q^*}{2} + f \times \frac{b}{q^*} \tag{2.20}$$

Consider an example. Demand for an LED TV is 3200 units per quarter. The fixed costs per order are \$2500. Annual holding costs per LED TV are \$80. Calculate the number of LED TVs that the store manager should order per refill.

- **Solution**

Annual demand: $b = 3200 \times 4 = 12{,}800$ units; ordering cost/order: $f = \$2500$.

Holding cost per unit per year: $c = \$80$.

Using the EOQ formula, the optimal order quantity is

$$q^* = \sqrt{\frac{2 \times 12{,}800 \times 2500}{80}} = 895$$

To minimize total costs, the store manager orders 895 LED TVs per refill. The cycle inventory is the average resulting inventory and is calculated as follows: cycle inventory = 895/2 ≈ 448 units.

For an order size of $q^* = 895$, the store manager evaluates the following:
- Number of orders per year: $N = 12{,}800/895 = 14.3$.
- Expected time between orders: $T = 365/14.3 = 25.5$ days.
- Total cost: $\text{TC} = 80 \times 895/2 + 2500 \times 12{,}800/895 = \$71{,}554$.
- Total cost for 1 year is $71,554 for 14.3 orders with 895 LED TVs in each order.

In applying EOQ computation results and inventory control policies, the practical reality must be considered. For example, in some cases, suppliers fix the dates for new orders (e.g., each Friday) or allow orders at fixed quantities (e.g., 50, 150, or 300 units) only.

### 2.2.2 Reorder Point

While the EOQ model answers the "how much" question, the *reorder point* (ROP) tells "when" to order. ROP is introduced to take into account the so-called *lead time*, i.e., the time between placement and receipt of an order (see ◘ Fig. 2.5).

With the assumption of constant demand and a set lead time, ROP is calculated as in Eq. (2.21):

$$\text{ROP} = d \times L, \tag{2.21}$$

where $d$ is the daily demand and $L$ is the lead time.

*Note:* In ROP, we consider daily demand. If we are given annual demand, it should be divided into the corresponding number of working days in a year.

Consider an example. A company experiences an annual demand of 8500 cheese knives per year (250 working days). Lead time for an order is five working days. Calculate the ROP and explain its meaning.

- **Solution**
  1. Daily demand = 8500/250 = 34 units
  2. ROP = $34 \times 5$ = 170 units

If our inventory reaches 170 cheese knives, then we have to place a new order.

◘ **Fig. 2.5** Reorder point

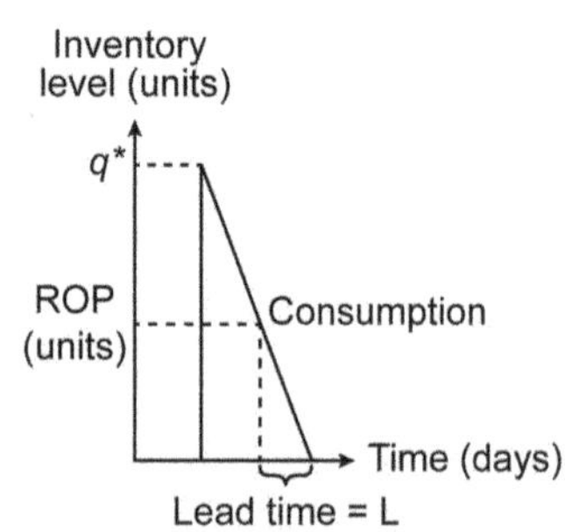

### 2.2.3 Service Levels and Safety Stock

Uncertainty in demand makes it necessary to maintain a certain customer service level or *level of product availability* to avoid stock-outs. The level of product availability is the fraction of demand that is served on time from a product held in inventory. A high level of product availability provides a high level of *responsiveness* but increases costs because much inventory is held but rarely used. In contrast, a low level of product availability lowers inventory holding cost but results in a higher fraction of customers who are not served on time. The basic trade-off when determining the level of product availability is between the cost of inventory to increase product availability and the loss from not serving customers on time.

For example, a 0.05 probability of stock-out corresponds to a 95% service level (see ◘ Fig. 2.6).

In practice, different *service level* estimations are used. For example, *anyLogistix* software uses two service level indicators (Ivanov, 2020):

- *Service level* is the ratio of orders delivered directly from a warehouse using inventory on hand to total number of orders (this indicator is also known as the *fill rate*).
- The *expected lead time (ELT) service level* is the ratio of orders delivered to a customer within expected lead time (i.e., on-time orders) to total number of orders received from this customer (this indicator is also known as the *on-time delivery*).

In a situation of demand uncertainty, *safety stock* is introduced with the objective to increase product availability even in the case of demand fluctuations (◘ Fig. 2.7).

In ◘ Fig. 2.7, the ROP is increased by safety stock. The question is: *How much safety stock should we plan to find a right balance between the inventory investment and customer satisfaction?* Technically, the safety stock computation is based on the desired service level and demand volatility. In order to compute safety stock subject to a desired service level, Eq. (2.22) is used:

$$\mathrm{ss} = z \times \sigma_{\mathrm{dLT}}, \tag{2.22}$$

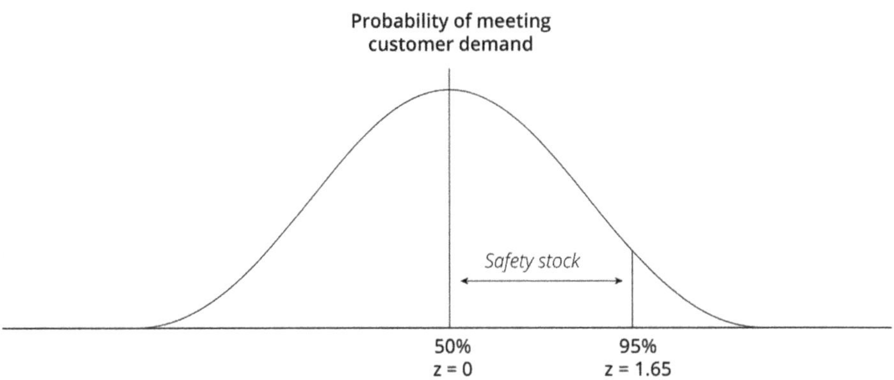


◘ **Fig. 2.6** Interrelation of demand distribution, ROP, service level, and safety stock. (Adapted from Heizer & Render, 2013)

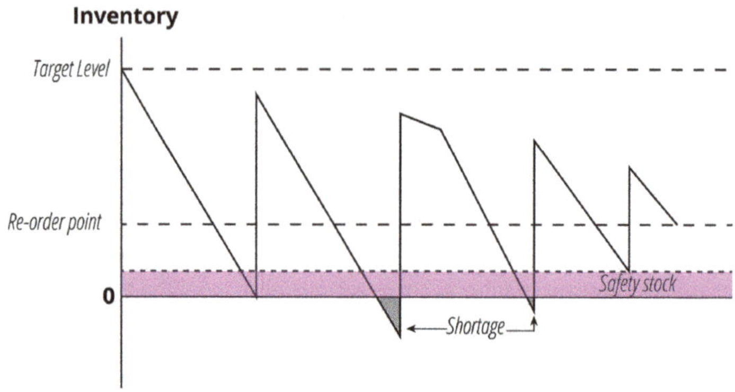


**Fig. 2.7** ROP with safety stock (Ivanov et al., 2021)

where ss is the safety stock, $\sigma_{\mathrm{dLT}}$ is the standard deviation of demand during lead time, and $z$ is the number of standard deviations.

Demand deviation can be gleaned, e.g., from the analysis of demand forecasts and actual sales in the past. For example, $\sigma = 1.25\mathrm{MAD}$ is a typical value. The $Z$-value can easily be determined using the table of normal distribution (see, e.g., Ivanov et al., 2021, p. 406). For example, $z = 1.65$ for a service level of 95%, $z = 2.33$ for a service level of 99%, and $z = 1.28$ for a service level of 90%. If standard deviation of demand during lead time is 10, then safety stocks equal $1.65 \times 10 = 16.5$; $2.33 \times 10 = 23.3$; and $1.28 \times 10 = 12.8$. We can observe that the increase of service level from 90% to 99% results in a doubling of inventory.

The inclusion of safety stock changes the calculation of ROP (see Eq. 2.23):

$$\mathrm{ROP} = \bar{d} \times L + \mathrm{ss}, \tag{2.23}$$

where $\bar{d}$ is the average daily demand.

In order to calculate ROP, four situations are possible:

- Demand is assumed to be normally distributed during the lead time.
- Daily distribution of demand is given (i.e., demand is variable), and lead time is constant.
- Daily demand is constant, and lead time is variable.
- Both demand and lead time are variable.

In order to calculate ROP, if demand is assumed to be normally distributed during the lead time, formula (2.24) can be used:

$$\mathrm{ROP} = \bar{d} \times L + z \times \sigma_{\mathrm{dLT}} \tag{2.24}$$

If daily distribution of demand is given (i.e., demand is variable) and lead time is constant, formula (2.25) can be used:

$$\mathrm{ROP} = \bar{d} \times L + z \times \sigma_d \times \sqrt{L} \tag{2.25}$$

If daily demand is constant and lead time is variable, formula (2.26) can be used:

$$\mathrm{ROP} = \bar{d} \times L + z \times \bar{d} \times \sigma_L \quad (2.26)$$

Equation (2.26) nicely illustrates the importance of reducing lead time variability. We can observe that lead time variability reduction directly influences safety stock levels. This observation depicts the integration of supplier selection, contracting, and inventory management decisions.

Finally, if both demand and lead time are variable, formula (2.27) can be used:

$$\mathrm{ROP} = \bar{d} \times L + z\sqrt{L \times \sigma_d^2 + \bar{d}^2 \times \sigma_L^2} \quad (2.27)$$

### 2.2.4 Single Period Systems ("Newsvendor Problem")

The newsvendor problem is a mathematical model for calculating the optimal inventory level for one single period. It is called the newsboy or newsvendor problem. A newspaper vendor who must decide every day how many daily newspapers he wants to stock for the next day is faced by uncertain demand and the knowledge that unsold copies will be almost worthless next day.

The newsvendor problem is characterized by the following conditions:

- Fixed price for each unit
- Perishable product
- Uncertain demand
- No additional delivery in period $t$
- Short purchase time

Consider the following notations for single period model:

- $c$ is the purchase price
- $r$ is the retail price
- $v$ is the salvage price
- $c_o$ is the overage cost
- $c_u$ is the underage cost
- $z$ is the number of standard deviations
- $\sigma$ is the standard deviation of demand
- $\mu$ is the expectation of demand
- $f_{01}(z)$ is the probability density function
- $S$ is the order quantity
- $S^*$ is the optimal order quantity
- $Z(S^*)$ is the expected cost for optimal order quantity
- $\prod(S^*)$ is the expected profit for optimal order quantity

To calculate the *overage* and *underage costs*, we can use Eqs. (2.28) and (2.29):

$$c_o = c - v \quad (2.28)$$

$$c_u = r - c \quad (2.29)$$

2

Then, we use the *critical ratio* (CR) to find the $z$-value from the table of normal distribution (Eq. 2.30):

$$\text{CR} = \frac{c_\text{u}}{c_\text{u} + c_\text{o}}; \quad F(\text{CR}) = z \tag{2.30}$$

For example, $\text{CR} = \frac{0.75}{0.75 + 0.25}; F(0.75) = 0.68 = z$.

Next step is the calculation of $S^*$ according to Eq. (2.31):

$$S^* = \mu + z \times \sigma \tag{2.31}$$

In order to calculate the expected cost and profit for optimal order quantity, Eqs. (2.32) and (2.33) can be used:

$$Z(S^*) = (c_\text{o} + c_\text{u}) \times f_{01}(z) \times \sigma \tag{2.32}$$

$$\Pi(S^*) = c_\text{u} \times (\mu - Z(S^*)) \tag{2.33}$$

Consider an example. Coff & Co., a coffee shop, sells vegan chocolate croissants. They purchase the croissants from a small bakery at the end of the street. The bakery sells chocolate croissants to Coff & Co. for $0.70 each. Coff & Co. sells them for $2.40 to their customers. Unsold croissants can be returned to the bakery for $0.15 each. On the basis of the last few months, Coff & Co. expects a normal distributed demand for chocolate croissants. Expectation of demand is 14 croissants per day with a standard deviation of four per day. Calculate the optimal order quantity and expected costs and profit for the chocolate croissants.

- **Solution**

$$\text{Overage cost}: c_\text{o} = c - v = 0.70 - 0.15 = \$0.55$$

$$\text{Underage cost}: c_\text{u} = 2.40 - 0.70 = \$1.70$$

$$\text{CR} = \frac{1.70}{1.70 + 0.55} = 0.75; \quad F(0.7556) = 0.7 = z$$

$S^* = 14 + 0.7 \times 4 = 17$ units; therefore, Coff & Co. should order 17 vegan chocolate croissants per day.

$Z(S^*) = (0.55 + 1.70) \times f_{01}(0.7) \times 4 = \$2.81$/day; so expected cost is $2.81/day.

$\Pi(S^*) = 1.70 \times (14 - 2.81) = \$20.99$/day; so expected profit is $20.99/day.

*Note*: $f_{01}(z)$ value can be taken from the full version of a normal distribution table; see, e.g., Thonemann (2010, pp. 272–275).

### 2.2.5 Periodic Inventory Control

*Inventory control policy* is a managerial procedure that helps define how much and when to order. The review may happen periodically (e.g., at the end of a month) or continuously (i.e., tracking each item and updating inventory levels each time an

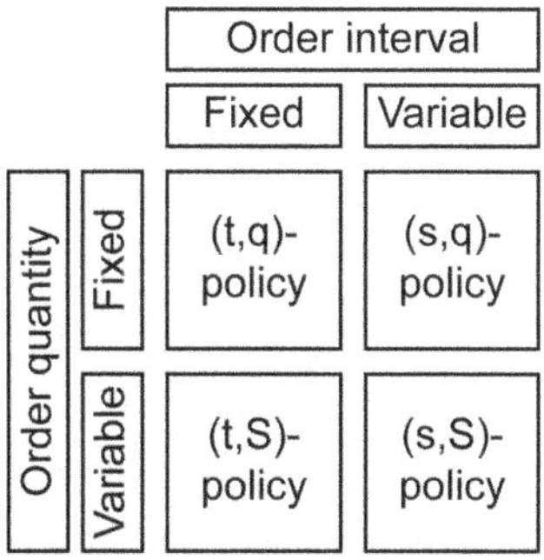


**Fig. 2.8** Inventory control policies (Ivanov et al., 2021)

item is removed from inventory (I)). Four parameters are important in the setting up of inventory control policies:

- *t* is the *replenishment interval*
- *q* is the *order quantity*
- *s* is the *reorder point*
- *S* is the *target inventory level*

Since order quantity and replenishment intervals may be both fixed and variable, four basic inventory control policies can be classified (see Fig. 2.8).

If the period between two orders is always the same, we talk about *periodic review inventory control.* If the point of time of the next replenishment depends on the ROP, we talk about the ROP method of stock control or a *continuous review inventory control.* The abovementioned four parameters can be fixed or changed (adjusted) in dynamics according to changes in demand and supply.

- **Policy 1: *t, q***
- *t*: fixed time between two orders
- *q*: fixed order quantity

In (*t*, *q*) policy, a fixed amount (*q*) is ordered for a fixed period of time (*t*) (see Fig. 2.9).

(*t*, *q*) is a simple policy for handling the ordering process. This policy opens possibilities to further automatic control, which improves quality and saves resources, such as labor, energy, or materials. However, the order timing and quantities following the (*t*, *q*) policy will be contractually specified. This implies that the replenishment is simple but also inflexible. Should uncertainty or fluctuation in demand exist, this policy cannot be adjusted. In addition, shortage or overstocking makes the (*t*, *q*) policy an unattractive tool for many companies. Thus, it is recommended to implement this policy under constant demand.

- **Policy 2: *t, S***
- *T*: fixed time between two orders
- *Q*: variable order quantity to stock up to the target level *S*

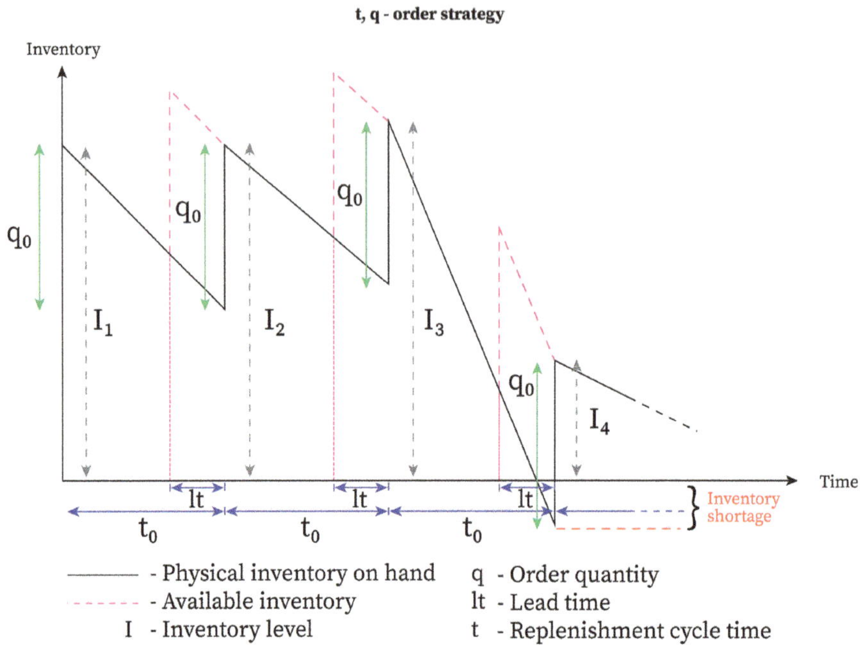


**Fig. 2.9** ($t$, $q$) inventory control policy (Ivanov et al., 2021)

In the ($t$, $S$) policy, the order quantity ($q$) is variable, and $q$ is placed at a fixed time ($t$). We need to order a certain amount of inventory to reach the desired quantity $S$ subject to lead time (lt). Order quantity is calculated as the target inventory level $S$—order level (see Fig. 2.10).

This policy avoids excessive inventory, which cannot be used for any other purpose and thus involves opportunity costs. The model is easy to use for control of orders. The unbroken line depicts physical inventory on-hand. The dashed line represents available inventory which is measured as inventory-on-hand plus planned future receipts minus open orders (shortage). However, the physical control of the inventory could be so expensive that the exact count is only performed once a month, for example. In certain cases, the ($t$, $S$) policy can lead to relatively high capital commitment because of the high average inventory. This policy also implies high ordering costs because we might not place a large order on the fixed day. At the same time, we might need to wait too long to fulfil our target inventory, and thus a shortage can occur. The ($t$, $S$) policy is recommended for use in companies with cycled replenishment.

### 2.2.6 Continuous Inventory Control

- **Policy 3: *s*, *q***
  - $t$: variable time between two orders
  - $q$: fixed order quantity

This model operates when order quantity ($q$) is fixed and the interval ($t$) between orders can vary. In this case, the order point ($s$) is defined as ROP (see Fig. 2.11).

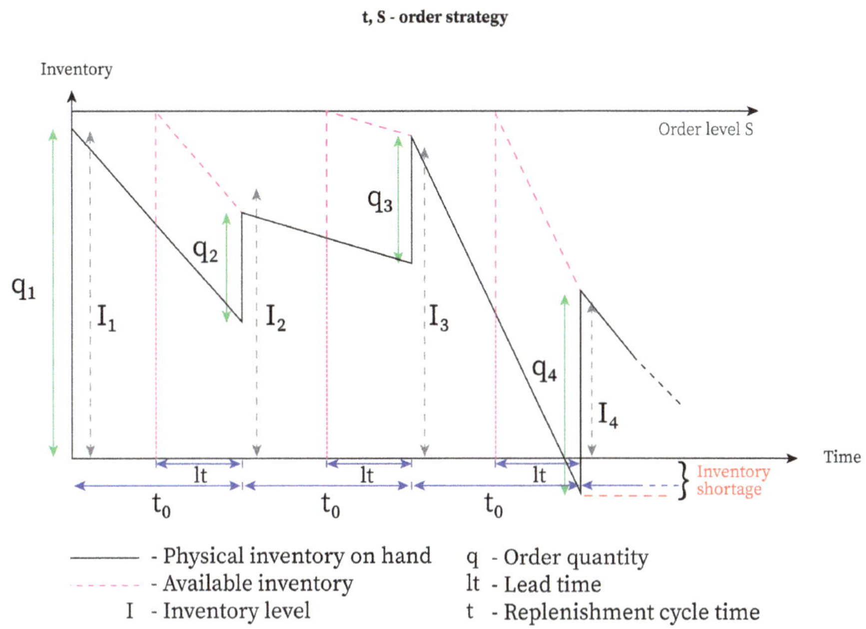


**Fig. 2.10** ($t$, $S$) inventory control policy (Ivanov et al., 2021)

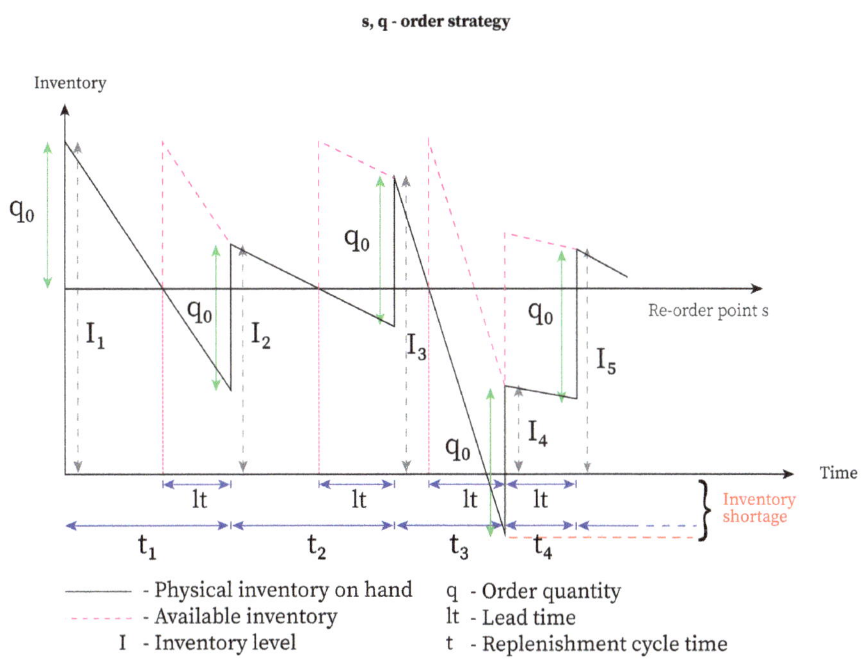


**Fig. 2.11** ($s$, $q$) inventory control policy

2

Every order arrives to replenish inventory after a lead time. The lead time is assumed to be known and constant. The only uncertainty is associated with demand. In the following analysis, one should be most concerned with the possibility of shortage during an order cycle, that is, when the inventory level falls below zero. This is also called a stock-out event. Every time we extract inventory, we compare what is left with $s$.

Note: for calculating the ROP, we use Eq. (2.34):

$$\text{ROP} = d \times (T + L) + z \times \sigma \times \sqrt{(T + L)} \tag{2.34}$$

If the stock level is less than $s$, then we place an order at the rate of $q$. Similar to the ($t$, $q$) policy, in the ($s$, $q$) policy, $q$ also refers to the optimal order quantity. The policy ($s$, $q$) results in shorter time between orders if there is inventory shortage. Because of its simple operation and full control over results, this policy is widely used in organizations. An advantage of the model is that it considers demand fluctuations. Disadvantages lie in regular inventory control.

- **Policy 4: *s*, *S***
  - $t$: variable time between two orders
  - $q$: variable order quantity between the order level $S$ and ROP $s$

This strategy is used to define the drop of order quantity $s$ after every inventory usage. Should this be the case, a manager should refill inventory to raise the inventory position to the level $S$, which is a desirable property (see ◘ Fig. 2.12).

Therefore, both order quantity and time interval between orders are variable. This system can handle any level of demand and at any time and includes demand fluctuations in planning. Target level is calculated as Eq. (2.35):

$$S = (\text{ROP} + q) \tag{2.35}$$

Order policy ($s$, $S$) avoids an excessive level of inventory and ensures that the business has the right goods on hand to avoid stock-outs. However, this policy requires much effort and high control. It is used in industrial and commercial areas of business, given the fact that flexible order quantity is possible and a target quantity can be predetermined.

Consider an example:

- Demand per day ($d$)
  100 units
- Standard daily deviation of demand ($\sigma$)
  20 units
- Annual holding costs ($h$)
  $10 per unit
- Fixed ordering costs ($f$)
  $100 per order
- Order interval ($T$)
  4 days
- Lead time ($L$)
  2 days

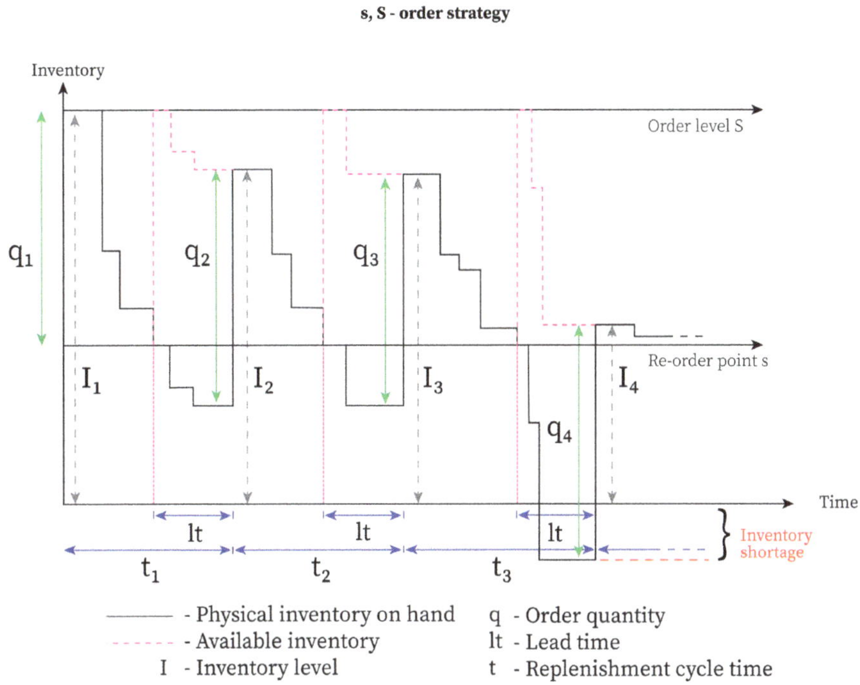


**Fig. 2.12** ($s$, $S$) inventory control policy

- Service level
  95%

- **Solution**
  1. Find $z$-value for 95% service level; we get 1.65

$$\mathrm{ss} = z \times \sigma \times \sqrt{(T+L)} = 1.65 \times 20 \times \sqrt{4+2} = 81 \text{ units}$$

$$\mathrm{ROP} = 100 \times (4+2) + 81 = 681 \text{ units}$$

  2. $S = \mathrm{ROP} + q;\ q^* = \sqrt{\dfrac{2 \times 36{,}500 \times 100}{10}} = 855 \text{ units};\ S = 681 + 855 = 1536 \text{ units}$
  3. The policy is (681; 1536)

## 2.3 Production Planning

### 2.3.1 Problem Statement

Consider an example. A company produces two types of memory cards: A and B. The selling prices are \$80 and \$100, respectively. Variable production costs are

\$68 and \$85, respectively. Maximum demand forecast is 100 units for A and 150 units for B. All the products are produced at a machine with a total capacity of 1000 units. Production of product A seizes 5 capacity units. Production of product B seizes 4 capacity units. What is the optimal product mix in a production plan in terms of profit maximization?

### 2.3.2 Linear Programming Method

*Linear programing* (LP) is a widely used quantitative analysis tool to support decisions about allocating available resources subject to demand in order to maximize profit or minimize costs (e.g., in the transportation problem). The following is characteristic for LP models and describes its requirements and assumptions:

- An objective function that describes the management goal (e.g., profit maximization)
- Constraints that limit the degree to which we can pursue our objective
- Alternative ways for resource allocation
- Both objective and constraints must be expressed as linear functions

Let us formulate the LP models for the case described above. Our objective is to maximize the profit. We can state the following general objective function (Eq. 2.36):

$$Z = \sum_{i=1}^{n} p_i \times x_i \rightarrow \quad \text{or} \quad Z = p_A \times x_A + p_B \times x_B \rightarrow \max! \tag{2.36}$$

where $p_A$ and $p_B$ and $x_A$ and $x_B$ are the profits and quantities of products A and B, respectively.

Since the price for A is \$80 and the variable cost is \$68, $p_A$ = \$12. Similarly, $p_B$ = \$15. We can rewrite the objective function as follows (Eq. 2.37):

$$Z = 12x_A + 15x_B \rightarrow \max! \tag{2.37}$$

What is the maximal possible value of the objective function? Without having any constraints, the profit can be unlimited if we decide to produce an unlimited quantity of A and B. However, we have limited demand and capacity and have therefore to consider a constraint system. The solution to this constraint system will allow us to find quantities $x_A$ and $x_B$, such that the $Z$-value will be the maximal possible profit under the considered constraints, prices, and costs.

The constraint system in general form can be written as follows (Eqs. 2.38–2.40):

$$\text{Production capacity}: \sum_{i=1}^{n} c_i \times x_i \le C_{\text{total}} \quad \text{or} \quad C_A \times x_A + C_B \times x_B \le C_{\text{total}} \tag{2.38}$$

$$\text{Demand}: x_A \le D_A; \quad x_B \le D_B \tag{2.39}$$

$$\text{Non-negativity constraint}: x_A, x_B \ge 0 \tag{2.40}$$

where

- $C_A$ and $C_B$ are the number of capacity units required for production of one unit of A and B, respectively
- $C_{total}$ is the overall available capacity
- $D_A$ and $D_B$ are the demands for A and B, respectively.

The constraint system for the case with memory cards can be therefore written in the following way (Eqs. 2.41–2.43):

$$\text{Production capacity}: 5x_A + 4x_B \leq 1000 \tag{2.41}$$

$$\text{Demand}: x_A \leq 100; \quad x_B \leq 150 \tag{2.42}$$

$$\text{Non negativity constraint}: x_A, x_B \geq 0 \tag{2.43}$$

### 2.3.3 Application Example in Excel Solver

In order to solve the LP model, we use Excel Solver® that needs to be activated and is placed in the "Data" in the Analysis group on the Data tab.

In the first step, we set up the initial data about prices, costs, demand, and capacity constraints (◘ Fig. 2.13).

In cell B9, we define the objective function following Eq. (2.37). In cells B6 and C6, we set up demand constraints (2.42). We also prepare definition of the capacity constraint (2.41) in cells B8, C8, and D8.

Second, the objective function and the constraints are defined in Excel Solver (◘ Fig. 2.14).

First, we define that the value of the objective function will be written into cell B9. Second, we set up demand constraints specifying that the values in cells B11

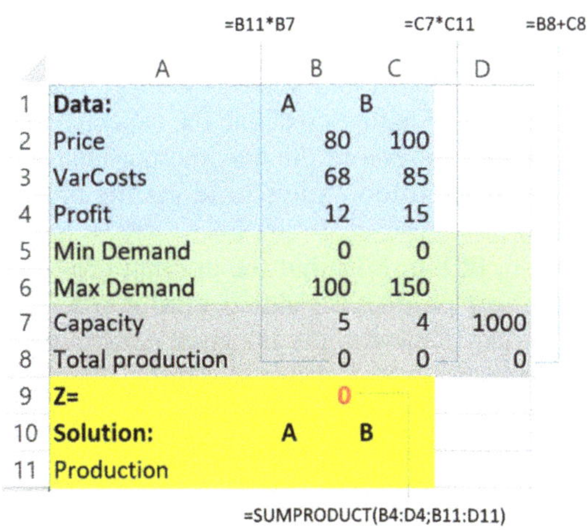


◘ **Fig. 2.13** Input data

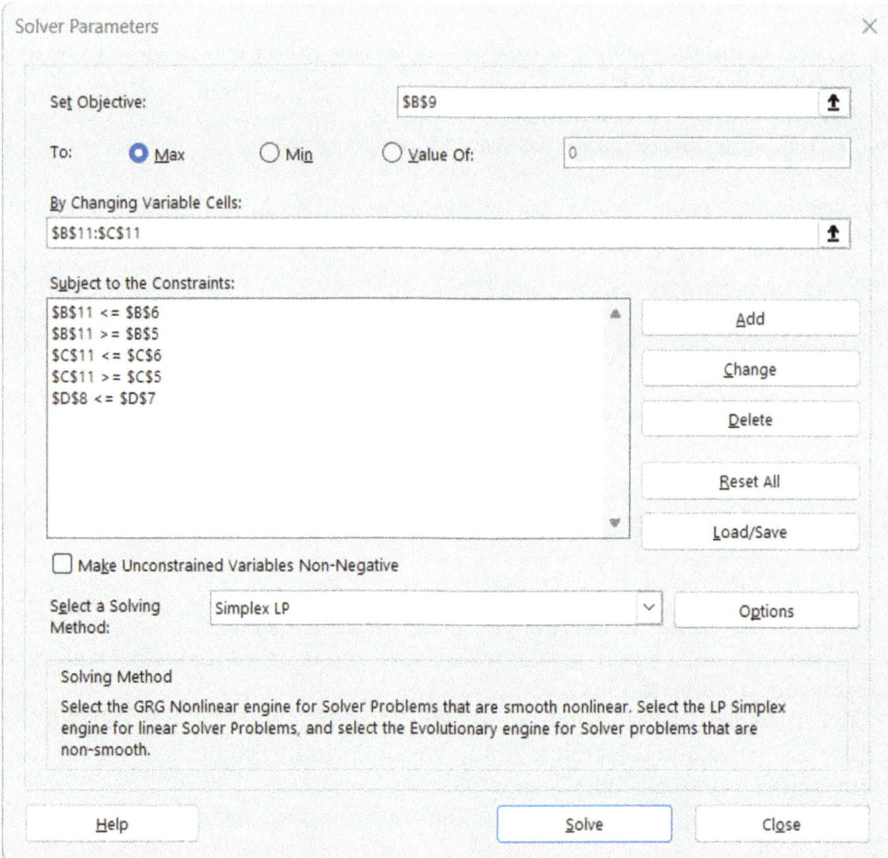


**Fig. 2.14** Solver model definition

and C11 (i.e., the decision variables $x_A$ and $x_B$) should not exceed maximum forecasted demand (cells B6 and C6) and should be more than minimum demand thresholds (cells B5 and C5). Finally, we define the capacity constraint saying that the value in cell D8 should not exceed the maximum available capacity defined in cell D7. Then, we press the button "Solve" and get the optimal solution to our problem (Fig. 2.15).

It can be observed in Fig. 2.15 that the optimal product mix is 80 units of product A and 150 units of product B yielding a profit of $3210. This is the maximum possible profit we can achieve under the given constraints. Indeed, product B yields the higher profit as product A. It is natural that the algorithm proposes to produce as many items of product B as possible according to the demand constraint (150 units). The remainder of the available capacity is then used to produce product A.

| Data: | A | B | |
|---|---|---|---|
| Price | 80 | 100 | |
| VarCosts | 68 | 85 | |
| Profit | 12 | 15 | |
| Min Demand | 0 | 0 | |
| Max Demand | 100 | 150 | |
| Capacity | 5 | 4 | 1000 |
| Total production | 400 | 600 | 1000 |
| **Total Profit Z=** | 3210 | | |
| **Solution:** | $X_A$ | $X_B$ | |
| Production | 80 | 150 | |

**Fig. 2.15** Solution of the LP model

## References

Heizer, J., & Render, B. (2013). *Operations management: Sustainability and supply chain management* (11th ed.). Pearson.

Ivanov, D. (2020). Supply chain optimization and simulation with anyLogistix. Retrieved from https://www.anylogistix.com/upload/alx-book.pdf; https://blog.hwr-berlin.de/Ivanov

Ivanov, D., Tsipoulanidis, A., & Schönberger, J. (2021). *Global supply chain and operations management: A decision-oriented introduction into the creation of value* (3rd ed.). Springer Nature.

## Further Reading

Axsater, S. (2015). *Inventory control* (3rd ed.). Springer.

Chase, C. W. (2013). *Demand-driven forecasting: A structured approach to forecasting* (2nd ed.). Wiley.

Chase, C. W. (2016). *Next generation demand management: People, process, analytics, and technology*. Wiley.

Choi, T. M., Wallace, S. W., & Wang, Y. (2018). Big data analytics in operations management. *Production and Operations Management, 27*(10), 1868–1883.

Chopra, S., & Meindl, P. (2015). *Supply chain management. Strategy, planning and operation* (6th ed.). Pearson.

Johnson, K., Lee, A. B. H., & Simchi-Levi, D. (2016). Analytics for an online retailer: Demand forecasting and Price optimization. *Manufacturing and Service Operations Management, 18*(1), 69–85.

Minner, S. (2000). *Strategic safety stocks in supply chains*. Springer.

Render, B., Stair, R. M., & Hanna, M. E. (2012). *Quantitative analysis for management* (11th ed.). Pearson.

Slack, N., Chambers, S., & Johnston, R. (2010). *Operations management* (6th ed.). Pearson.

Thonemann, U. (2010). *Operations management* (2nd ed.). Pearson.

Zipkin, P. H. (2000). *Foundations of inventory management*. McGraw-Hill.

# Discrete Event Simulation of Manufacturing Processes and Inventory Control

## Contents




D. Ivanov, *Introduction to Supply Chain Analytics*, Classroom Companion: Business,
https://doi.org/10.1007/978-3-031-51241-4_3

**Learning Objectives**

This chapter is devoted to the development of simulation models for manufacturing process and inventory control. First, we learn principles of queuing theory to understand the methodology of building discrete event-based process simulation models. Subsequently, AnyLogic software is introduced. We explain through illustrative case studies how to build simulation models and provide instructions for the development of process and inventory control policies using AnyLogic software.

To this end, our learning objectives for this chapter are as follows:

- Queuing theory and its application to manufacturing process control.
- Development of discrete event simulation models for manufacturing process and inventory control in AnyLogic software.
- Illustrative case studies.

## 3.1 Manufacturing Process Planning

### 3.1.1 Queuing Theory

*Queuing theory* is an important method for capacity planning in services and industries. Queuing theory helps to analyze *performance of a process* under temporal arrival of jobs or customers. Performance generally can be measured in different dimensions such as capacity, time, and cost. In ◘ Fig. 3.1, a simple queuing model (one-channel model) is presented.

A queuing model is comprised of the following elements:

- *Input* (e.g., patients)
- *Queue* (e.g., patients in the waiting room)
- *Processing station* (e.g., a doctor)
- *Output* (patients after the treatment)

In queuing theory, different options exist to describe these four elements. To describe the queues, we use the notation

$$a\,/\,b\,/\,c,$$

where

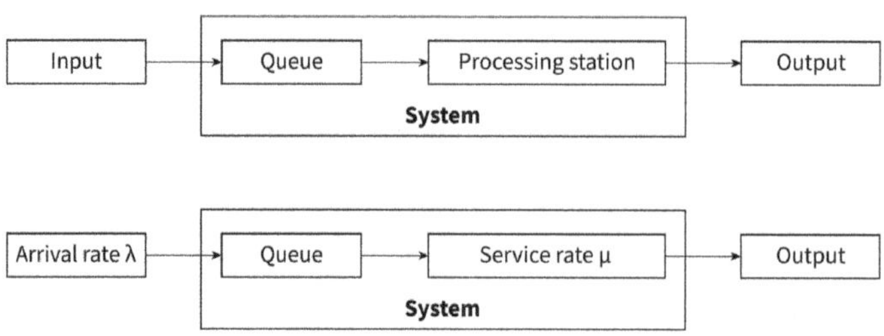


◘ **Fig. 3.1** Queuing model

*a* is the probability distribution of arrival rate, e.g.:
- *D* = degenerate distribution (deterministic)
- *M* = Markovian distribution (Poisson process)
- *G* = general distribution

*b* is the probability distribution of service time
- *D* = degenerate distribution
- *M* = exponential distribution
- *G* = general distribution

*c* is the number of servers (processing stations).

In regard to the number of servers, there are two basic models. A *one-channel model* considers only one queue and one server that processes the input. A *multiple-channel model* is involved with one queue and several servers or several queues and several servers.

Consider a system where there is no constraint in the queue and we follow the rule "first come first serve" for defining the processing sequence. We consider the so-called M/M/1 queue (◘ Fig. 3.1). We determine the mean number of arrivals per time period as $\lambda$ and the mean number of people or items served per time period (i.e., average service rate) as $\mu$. It means that the arrival rate is Markovian distributed, service time is exponentially distributed, and there is one server. We can find this queue, for example, in post offices or small supermarkets with only one cashier or in waiting rooms for only one doctor's surgery.

Consider the following formulas to understand how to analyze the performance of the queuing process. To determine the time between the arrivals of two customers, e.g., patients, we use Eq. (3.1):

$$\text{Estimated inter-arrival time} = \frac{1}{\lambda} \tag{3.1}$$

To analyze the service time, we use Eq. (3.2):

$$\text{Estimated service time} = \frac{1}{\mu} \tag{3.2}$$

To analyze the *utilization* factor for the system, we use Eq. (3.3):

$$\rho = \frac{\lambda}{\mu}; \quad \rho \leq 1 \tag{3.3}$$

To calculate the *probability* of more than *n*-units in the system, we can use Eq. (3.4):

$$P_n = \rho^n \cdot (1 - \rho) \tag{3.4}$$

To find out the *length of the queue* (e.g., the estimated number of patients in the waiting room), we use Eq. (3.5):

$$L_Q = \frac{\rho^2}{1-\rho} \tag{3.5}$$

The *length of the system* (e.g., the estimated number of patients in the waiting room *and* in the doctor's surgery) is determined with the help of Eq. (3.6):

$$L_s = L_Q + \rho = \frac{\rho}{1-\rho} \tag{3.6}$$

*Waiting time* in the queue (e.g., time spent in the waiting room) can be found using Eq. (3.7):

$$T_Q = \frac{\rho^2}{\lambda \cdot (1-\rho)} \tag{3.7}$$

*Waiting time* in the whole system (e.g., time spent in waiting room and doctor's surgery) is calculated with the help of Eq. (3.8):

$$T_s = T_Q + \frac{1}{\mu} = \frac{1}{\mu \cdot (1-\rho)} \tag{3.8}$$

In the case of an M/M/c queue, formulas are different. We can find M/M/c queues for example in a manufacturing process when a product can be processed at different machines. In the case of M/M/c queues, we have a Markovian distribution of arrival rate, an exponential distribution of service time, and *c* servers (>1).

Consider the following example for M/M/c queues. There are 40 jobs (i.e., customer orders) arriving per hour, and one workstation is able to process 12 jobs per hour. In total, there are 4 identical workstations capable to process the incoming jobs at some equal capacity rate. There is only one system queue.

To calculate the occupancy rate of the system, we use Eq. (3.9):

$$\rho = \frac{\lambda}{c \cdot \mu} = \frac{40}{4 \cdot 12} = 83.3\% \tag{3.9}$$

The total workstation capacity is occupied for 83.3% of the time.

To calculate the estimated number of jobs waiting in the queue, we should find the probability that the system is completely empty (Eq. 3.10):

$$P_0 = \frac{1}{\sum_{i=0}^{c-1} \frac{(c \cdot \rho)^i}{i!} + \frac{(c \cdot \rho)^c}{c!} \cdot \frac{1}{(1-\rho)}} \tag{3.10}$$

$$P_0 = \frac{1}{\frac{(4 \cdot 0.833)^0}{0!} + \frac{(4 \cdot 0.833)^1}{1!} + \frac{(4 \cdot 0.833)^2}{2!} + \frac{(4 \cdot 0.833)^3}{3!} + \frac{(4 \cdot 0.833)^4}{4!} + \frac{1}{(1-0.833)}}$$
$$= 2.14\%$$

The probability that the system is empty is 2.14%. Now it is possible to calculate the number of jobs in the queue (Eq. 3.11):

$$L_Q = \frac{c^c \cdot \rho^{c+1}}{c! \cdot (1-\rho)^2} \cdot P_0 = \frac{4^4 \cdot 0.833^{4+1}}{4! \cdot (1-0.833)^2} \cdot 0.0214 = 2.183 \qquad (3.11)$$

In case of four workstations and 40 estimated job arrivals per hour, 2.2 jobs are waiting in the queue.

The estimated number of jobs in the whole system (i.e., the WIP—work in progress) is calculated using Eq. (3.12):

$$L_s = L_Q + c \cdot \rho = 2.183 + 4 \cdot 0.833 = 5.515 \qquad (3.12)$$

On average, there are 5.5 jobs in the system.

Waiting time in the queue is defined using Eq. (3.13):

$$T_Q = \frac{L_Q}{\lambda} = \frac{2.183}{40} = 0.0546\ \text{h} = 3.276\ \text{min} \qquad (3.13)$$

Waiting time in the whole system is defined using Eq. (3.14):

$$T_s = T_Q + \frac{1}{\mu} = 0.0546 + \frac{1}{12} = 0.1379\ \text{h} = 8.274\ \text{min} \qquad (3.14)$$

As you can see, waiting time in the system is 8.3 min.

### 3.1.2 Little's Law

In 1961, John D.C. Little provided the mathematical evidence for the rule that puts the inventory $L$ (= average number of objects in a system) in relation to the throughput rate $\lambda$ (= average arrival rate of objects) and the waiting time $W$ (= average time an object spends in the system). Little's law says "that under steady-state conditions, the average number of items in a queuing system equals the average rate at which items arrive multiplied by the average time that an item spends in the system" (Little & Graves, 2008). Let:

- $L$ = average number of items in the queuing system (e.g., work-in-progress inventory in the system)
- $W$ = average waiting time in the system for an item
- $\lambda$ = average number of items arriving per unit time

The law can be represented as Eq. (3.15):

$$L = \lambda \cdot W \qquad (3.15)$$

*Little's law* demonstrates that when we reach a steady state, i.e., the system states are stationary and the arrival rate is smaller than the service rate (cf. ◘ Fig. 3.2), the average number of objects arriving in the system during a time period (e.g., 1 h or 1 day) is equal to the ones that are leaving the system during that time period. Elegant in its simplicity, Little's law represents one of the basics in capacity analysis. The following adapted example is related to the application of Little's law in a manufacturing plant described by Bozarth and Handfield (2013).

3

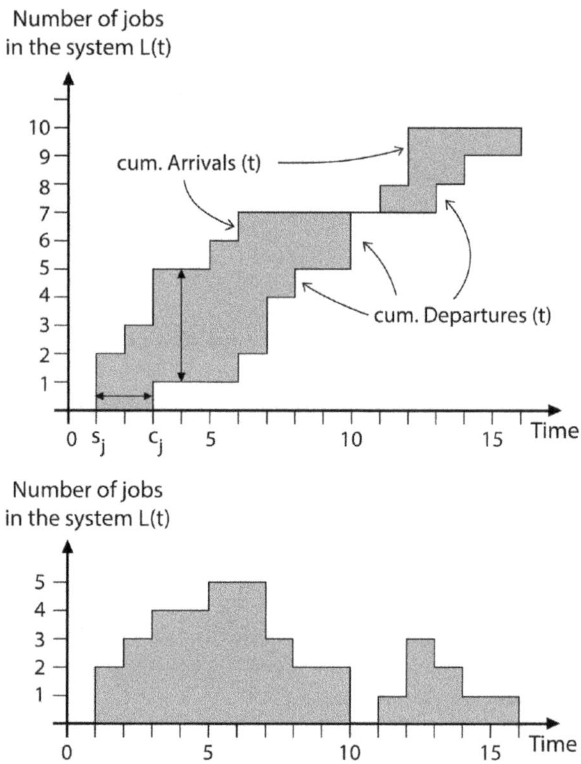


**Fig. 3.2** Inventory and lead time dynamics analysis (Ivanov et al., 2021)

Consider an application of the Little's law to inventory and lead time analysis in a manufacturing system (Fig. 3.2).

The upper part of Fig. 3.2 presents the dynamics of incoming and outgoing orders in a manufacturing system. At point of time 1, two orders arrive in the system and the number of jobs in the system, i.e., the inventory $L(t)$ is 2 at $t = 1$. At $t = 2$, one more order arrives, and first two orders are still processed. Therefore, $L(t) = 3$ at $t = 2$. At $t = 3$, two new orders arrive, and the first order exits the system. Therefore, $L(t) = 4$ at $t = 3$. This inventory dynamics is shown in the bottom part of Fig. 3.2. Moreover, the upper part of Fig. 3.3 shows lead time $(s_j - c_j)$ for each job, where $s_j$ is the completion time of the job and $c_j$ is the start time of job processing. For example, lead time for the job #1 is two time units.

### 3.1.3 Bottleneck Analysis/Theory of Constraints

In many practical settings, the queuing system is composed of multiple, subsequent stages, i.e., a manufacturing line that is comprised of several subsequent machines. In such systems, we frequently encounter so-called bottlenecks. A *bottleneck* is the

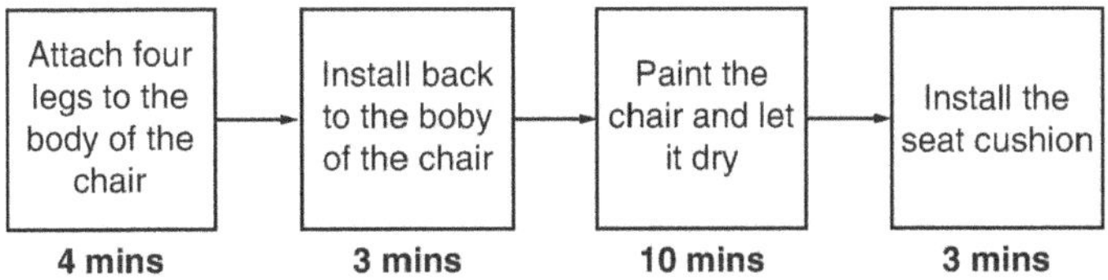


**Fig. 3.3** Process chain

part of a process with the longest unit processing time, and it determines the capacity of the system. Assume that there is a product (e.g., a chair) that requires four sequential production steps (see Fig. 3.3):

In this simplified production process, painting and drying operations have the longest duration of 10 min per chair. Putting simply, six chairs per hour can be painted, and thus the system capacity is six chairs as this is determined by the bottleneck. However, other stations in the process can be much more productive. After the identification of the bottleneck (step 1), the operations manager's tasks would be to seek opportunities as to how this bottleneck time could be reduced.

Another frequently used term for bottleneck is *constraint*, as it is the process that prevents the enterprise from achieving its objectives (Goldratt, 1990). If we want to produce ten chairs per hour, something needs to be done in order to limit the impact of that constraint or bottleneck; the performance manager needs to find a way (most probably with the help of a manufacturing engineer) to reduce the bottleneck time from 10 to just 6 min per chair. A possible solution might be to improve the spray process or to use a different type of paint or to increase the temperature for the paint to dry faster. That means a plan has to be created to mitigate the impact of the constraint and to reduce the bottleneck time (step 2). The operations manager and the manufacturing engineer are the dedicated resources that will need to work on the implementation of the identified measures (step 3).

Furthermore, one should try to better balance the work and maybe try to do some necessary preparatory work required for the painting, such as cleaning after the back of the body of the chair was installed. If this is also done, we are trying to spread the work from the bottleneck as well as to try to increase capability in related production processes (step 4). If this has been successfully implemented, the performance manager will start again to identify new constraints and bottlenecks as part of this continuous search for improvements (step 5).

### 3.1.4 Trade-Off Capacity Utilization vs. Lead Time

We now merge the previous discussions on the queuing theory and the Little's law (Fig. 3.4).

Four major events, i.e., arrival of a customer order in the manufacturing system queue, production start, production end, and customer order delivery, build the framework of the analysis. The time interval in between the first and the last event

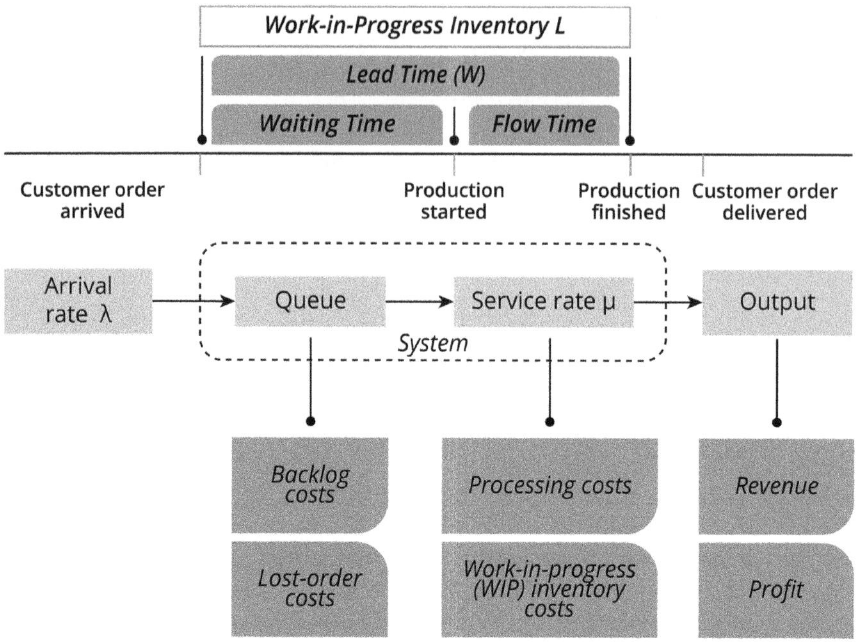


**Fig. 3.4** Major components of the waiting line and Little's law model

is called *lead time*. *Waiting time* is the time interval in between the first and the second event. *Flow time* is the time interval in between the second and the third event. During the waiting time, backlog costs and lost order costs occur. During the flow time, processing costs and WIP inventory costs are considered. The delivered orders generate revenues and profits.

**Important Observation**

A typical trade-off in process planning and control is between lead time and capacity utilization. Just imagine that you are in a supermarket on a Tuesday morning. Probably, you are the only person there at this time. So, you will not have to wait anywhere and will exit the supermarket in a short time. The time of your shopping will be short. But what about the supermarket? If you are the only customer, their capacity utilization is very low. The same effect can be encountered in manufacturing and service systems in regard to customer order lead time and resource capacity utilization. This typical *trade-off* is illustrated in Fig. 3.5.

If the number of orders or waiting customers would increase, this would lead to a higher capacity utilization. At the same time, lead time for each individual order or customer would increase due to longer waiting times. According to the trade-off, we cannot achieve a reduction in lead time and an increase in capacity utilization at the same time.

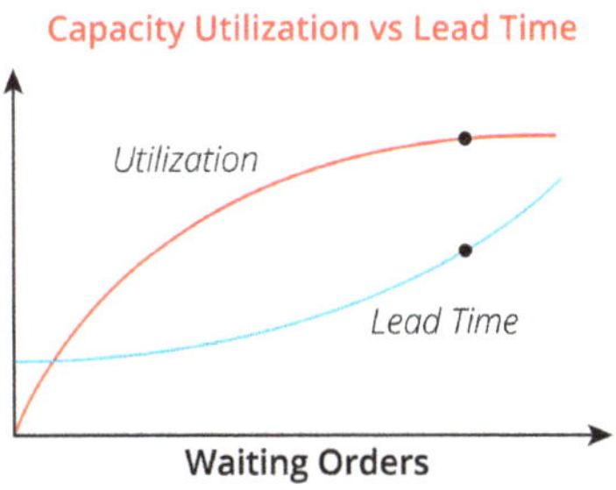


**Fig. 3.5** Trade-off: capacity utilization vs. lead time

**▶ Example**

Consider an example. Students at the Berlin School of Economics and Law arrive at the subway station Berliner Strasse, which is about a 10-min walk from the university. They can also take a bus to get to the school from the subway station. Assume that the buses run every 10 min, and every minute, one student comes to the bus stop. In this case, the waiting times and the lead times will be quite long, but the bus will take ten students and have good capacity utilization. Should the buses run every minute, the lead times would be very short (no waiting time), but the capacity utilization of each bus would be low. ◀

## 3.2 Discrete Event Process Simulation in AnyLogic

### 3.2.1 Problem Statement

In this section, we explain how to build *discrete event simulation* models in multi-method simulation software *AnyLogic*. These materials are based on the book "▶ Supply Chain and Operations Simulation with AnyLogic" by Ivanov et al. (2023), where more and detailed explanations can be found.

We consider a linear *manufacturing process* that contains four machines (Fig. 3.6):

Customer orders arrive at an uncertain rate at uncertain times. Two to five orders are expected every 2 h. Each order is processed in an identical sequence of the four machines M1–M4 (i.e., a flow shop) with a partially variable processing time at each machine. Processing times (in time units) follow the triangular distribution and are as shown in Table 3.1.

For example, processing of one order at the M1 can take 0.7 h (minimum duration), 1.3 h (maximum duration), or 1 h (most likely duration). Upon completion at the last machine, the order brings revenue of $30, but also incurs a $10 cost. Demand is backlogged such that if an incoming order cannot be processed due to insufficient line capacity, it waits. If the waiting time exceeds time limit of 14 days,

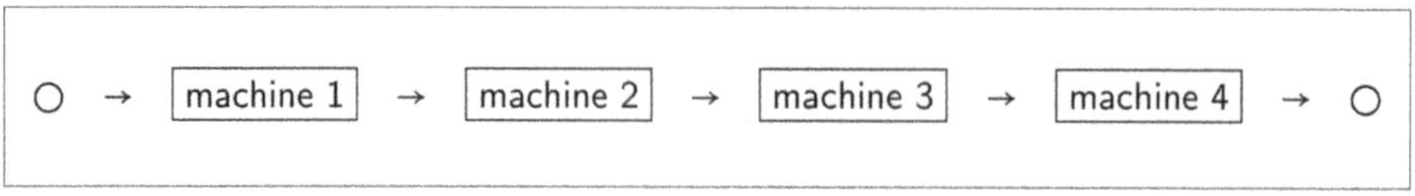


**Fig. 3.6** Manufacturing line

**Table 3.1** Processing times at machines

| M1 | M2 | M3 | M4 |
|---|---|---|---|
| triangularAV(1,0.3)<br>capacity 1 | triangularAV(1,0.3)<br>capacity 1 | triangularAV(2,0.3)<br>capacity 1 | triangularAV(1,1.3)<br>capacity 1 |

the customer withdraws the order, and cost of lost sales will be $4 per order. Each unprocessed backlog order creates a cost of $4 (this cost can be computed either at the end of the planning period or per unit time, e.g., for each week of waiting). Each partially processed (but not completed) order is called work-in-progress (WIP) and creates inventory costs of $0.05 per order per day to be computed similarly to the backlog cost. We consider a period of 100 h, i.e., we will have 100 arrivals of production batches, each of which will be comprised of 2–5 orders. In the simulation time, 1 s will equal 1 h of real life.

### 3.2.2 Development of Process Model in AnyLogic

#### Step 1: Create Process Model

In this step, we create a *process model* for our problem statement (Fig. 3.7).

The created model contains a source node where the customer orders arrive, four machines (M1–M4), the queues where the orders will wait if the capacity of respective subsequent steps is insufficient, and a sink node as the process ends. We also include four points to measure the elapsed times. Point Start1 measures the time of customer order arrival, point Start2 measures the time at which a customer order is dispatched into the manufacturing line, and points End1 and End2 measure the time at which the order is completed (End1 and End2 will measure the same moment in time). We refer to the duration End1 – Start1 as the process lead time and End2 – Start2 as the process flow time (compare with Fig. 3.4).

#### Step 2: Define Rules for Incoming Orders

In this step, we define the rules for the quantity and frequency of incoming orders (Fig. 3.8).

In the source properties, arrival can be defined by rates, interarrival time, etc. In the example in Fig. 3.8, the range of 2–5 orders will arrive with an interarrival time of 2 s. We limit our experiment to 100 arrivals.

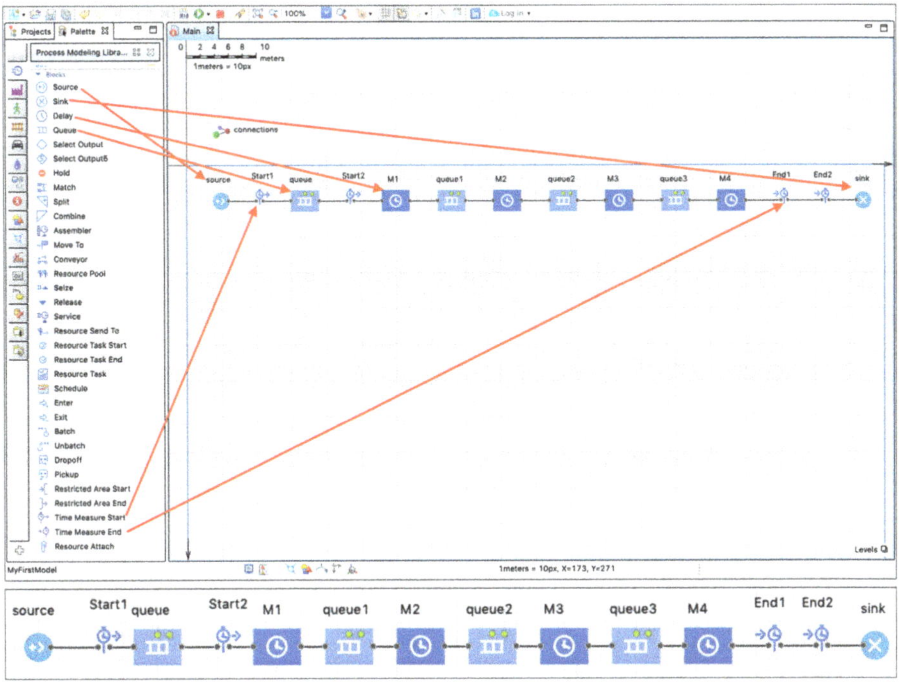


**Fig. 3.7** Process model

Next, the *push* vs. *pull* policy for the process can be selected in the source property field Forced pushing. By default, Forced pushing is enabled, and the source does not allow the orders to wait until they can be consumed by the next object. Deselect Forced pushing and choose the required behavior (destroy or wait in the queue) from the list Agents that cannot exit. This will imply a policy where orders are not pushed but rather pulled in the manufacturing line depending on the available capacity in correspondence to one-piece flow and *CONWIP* control techniques.

- **Step 3: Define Queue Rules**

In this step, we define the rules for the transition of customer orders into manufacturing line as well as internal transitions in the line with the help of setting the queue properties (Fig. 3.9).

First, the queue capacity can be limited or unlimited (maximum capacity). Second, dispatching rules can be set up as *First-In-First-Out (FIFO)*, priority-based (i.e., orders from the most important customers are dispatched first; this is activated by the Enable preemption option), agent comparison (i.e., preferences for certain orders), or *Last-In-First-Out (LIFO)*. The Enable exit on timeout option allows orders to abandon the queue (if an order is not dispatched during this time, it disappears). For example, if a customer order cannot be dispatched within

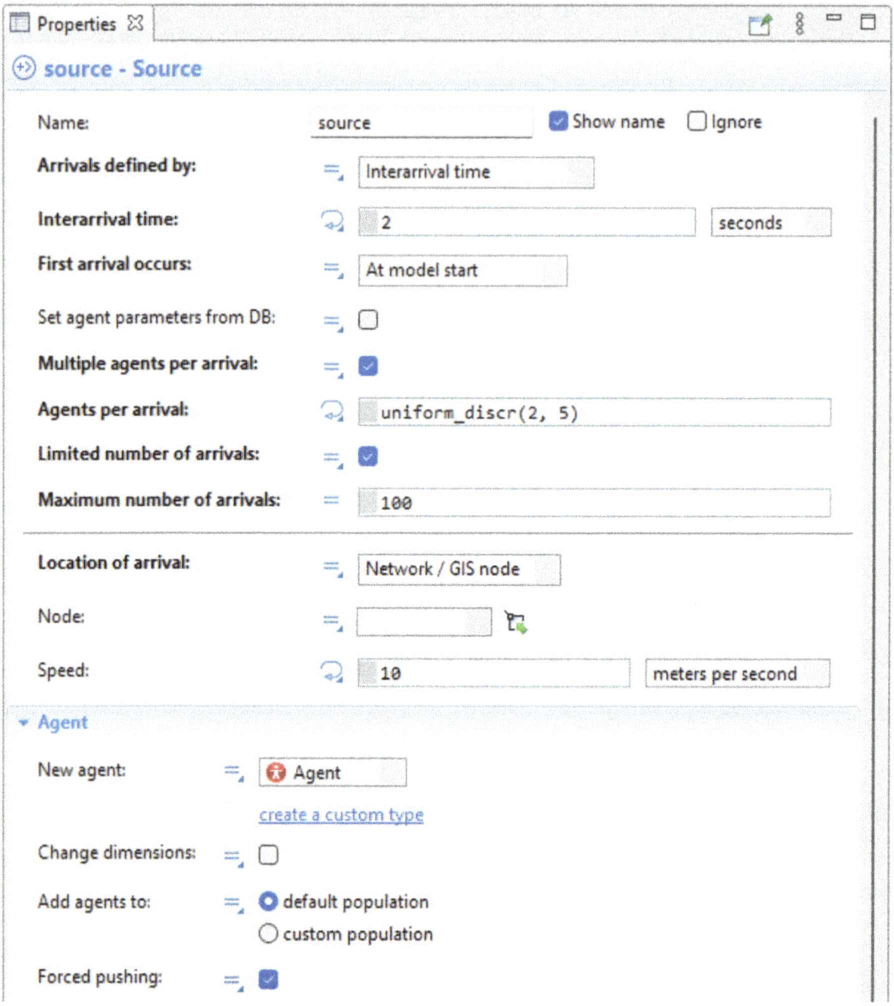


**Fig. 3.8** Defining incoming flow (input)

2 weeks, the customer will withdraw it and defect to our competitor. Such orders create opportunity costs that can be computed by comparing output at the sink vs. input at the source.

- **Step 4: Define Processing Rules**

In this step, we define the rules for order processing at the machines by setting the Delay properties (Fig. 3.10).

A processing time can be fixed or uncertain. In this case, we model processing times through a triangular distribution (see Table 3.1). For example, process-

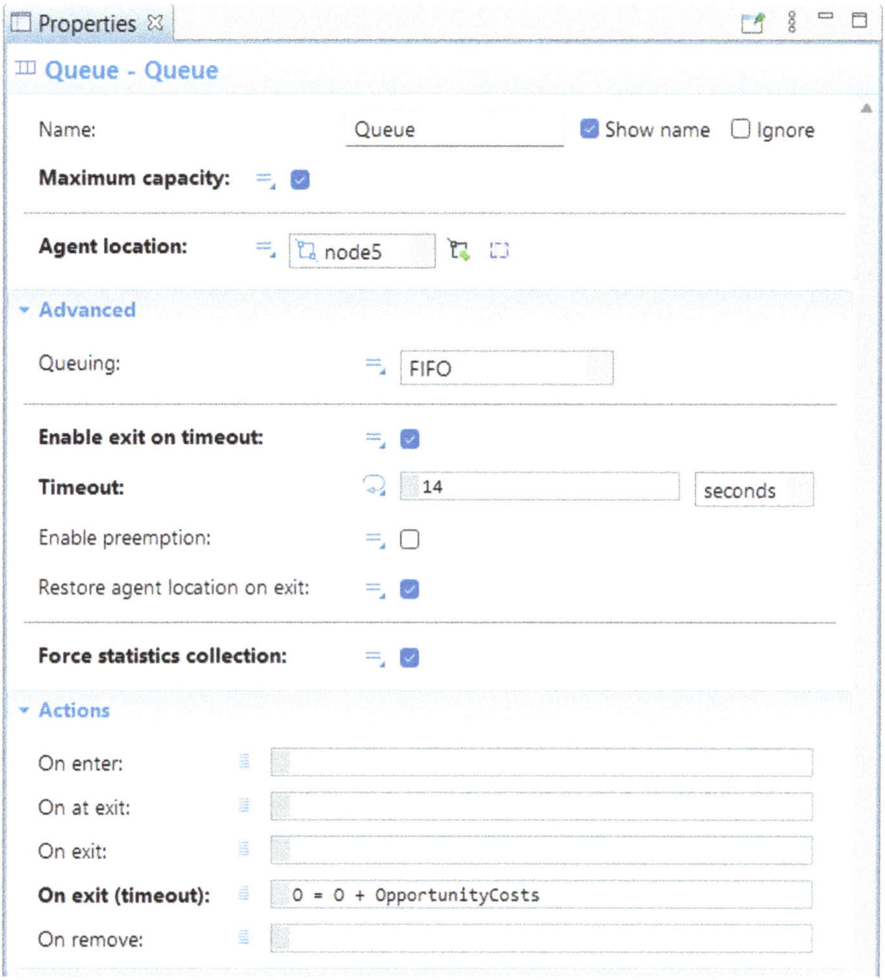


**Fig. 3.9** Defining queue rules

ing times at machine 1 range between 0.7 and 1.3 s, with 1 s being most likely. The capacity of the processing machine defines a maximum number of orders that can be processed at the machine simultaneously. In the advanced models, the processing stage can comprise not only machines but also different resources (e.g., a forklift). Such advanced descriptions can be implemented with the help of functions Resource, Seize, Release, or Service (see Ivanov et al., 2023). Repeat using the following triangular distributions for the Delay times at machines M2, M3, and M4.

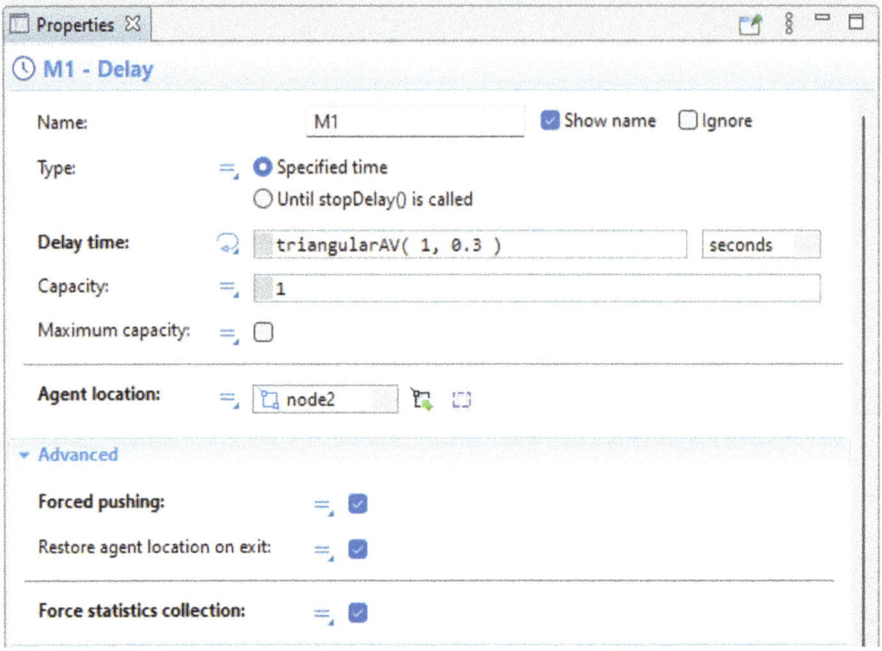


■ **Fig. 3.10** Defining processing rules

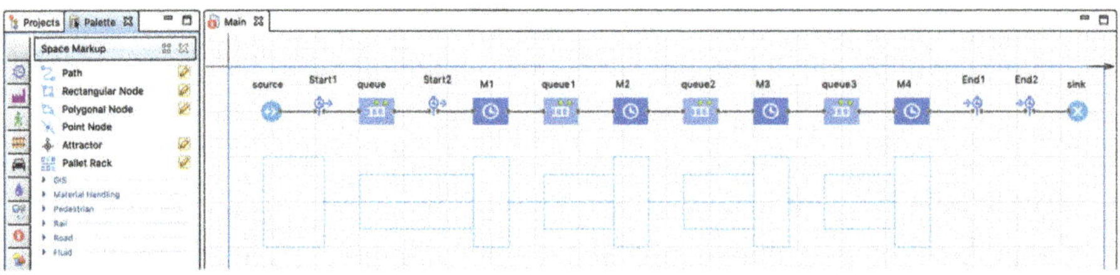


■ **Fig. 3.11** Process visualization

## 3.2.3 Development of Process Visualization in AnyLogic

- **Step 1: Creating Basic Animation and Running the Model**

One of the key simulation advantages is *process visualization*, which helps managers understand simulation results better and gives intuition about how systems are likely to behave. The range of different visualization levels is broad, from basic blocks to 3D animation.

We consider here a very simple procedure (■ Fig. 3.11).

Using the Space Markup palette, we draw nine Rectangular Nodes (the source node, four waiting areas, and four machines) by double-clicking on the Rectangular Node tool. We connect them with Paths (double-click the "Path" tool before dragging each path) as shown in ◘ Fig. 3.11. Open the Properties for each rectangular node, and give each a name. For example, ◘ Fig. 3.12 depicts labeling the first node "Supplier." In the properties of each process diagram element (i.e., Source, Queue, and Delay blocks), we assign the drawn Rectangular Nodes.

In ◘ Fig. 3.12, we assign the "Supplier" node to the Source block by setting the Location of arrival to "Network/GIS node" and the Node to "Supplier." Repeat this by opening the Properties of each Queue and Delay block, and choose the name of the corresponding Rectangular Node in Agent location.

Now we are ready to run the experiment by going to "Run Model" (◘ Fig. 3.13).

In the process diagram and visualization field, the operational process dynamics can be observed. In the "Projects – Simulation," experiment parameters can be set up, e.g., maximal experiment time of 100 s that corresponds to 100 days or hours of real life (◘ Fig. 3.14).

The process dynamics observation is very useful, but for managers, it is insufficient. A manager needs an evaluation of both financial (i.e., revenue, costs, profits)

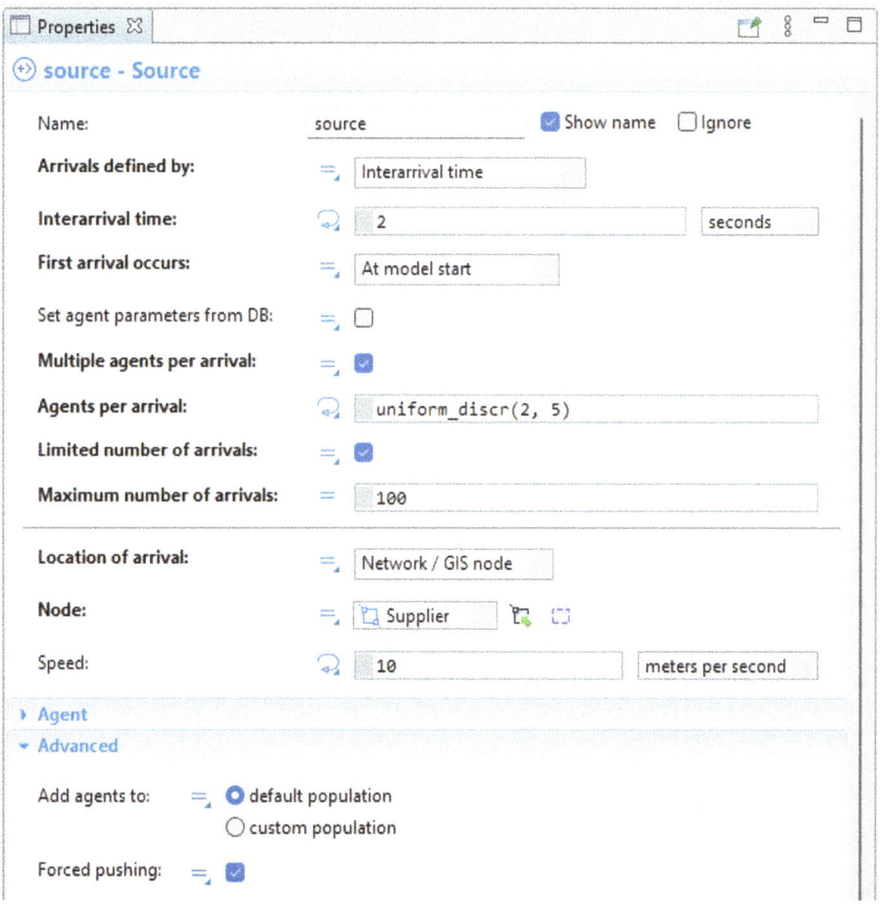


◘ **Fig. 3.12** Matching process diagram and visualization

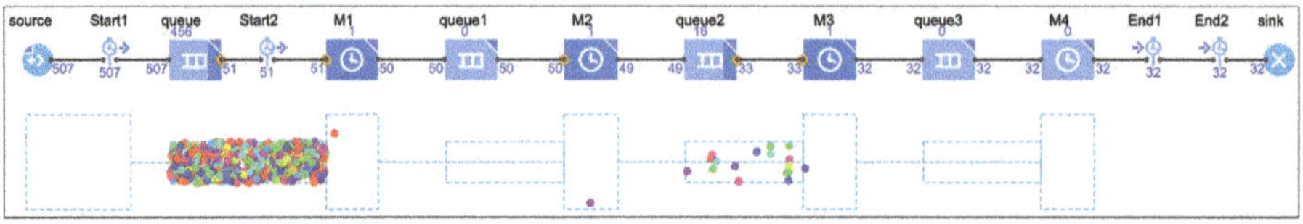




**Fig. 3.13** Experiment window

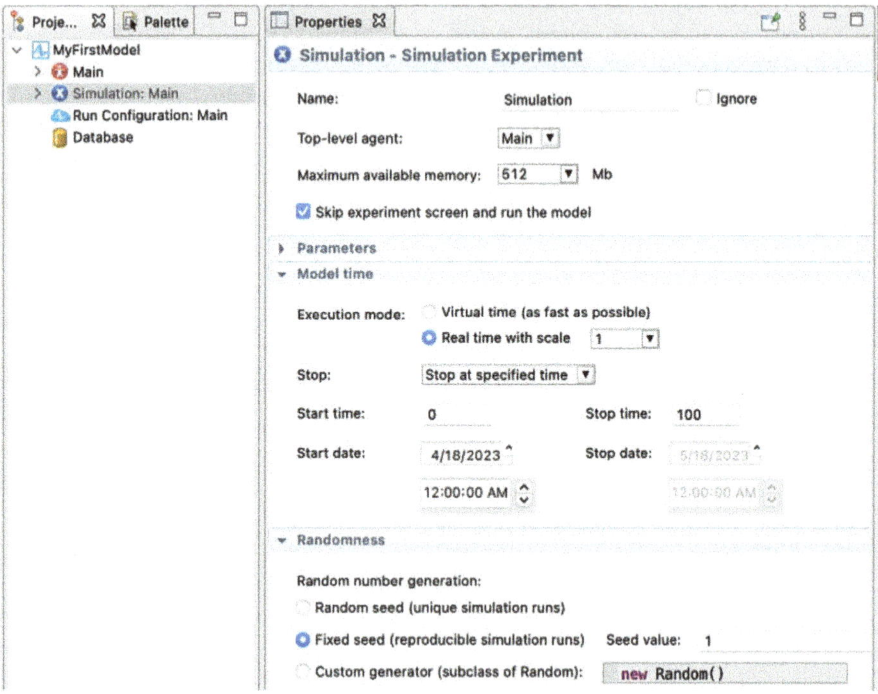


**Fig. 3.14** Experiment settings

and operational performance (i.e., lead time, flow time, capacity utilization, and backlog). In other words, in the last step prior to starting the experimental part, *a KPI cockpit* or *dashboard* needs to be created.

#### ▪ Step 2: Collecting Statistics and KPI Dashboard Design

In this step, we are going to create diagrams to analyze the following KPIs:

- Revenue, costs, profit
- Capacity utilization
- Lead time and flow time
- Backlog

Using the Analysis palette, we include various bar charts, stack charts, plots, time plots, and histograms as appropriate for the KPIs.

#### Step 2.1: Financial KPIs

To compute financial performance as revenue, costs, and profit, we should define some parameters and variables in "Palette → Agent" as shown in ◘ Fig. 3.15:

In the parameters "Costs," "Revenue," and "Backlog," we define the monetary estimations of manufacturing costs, revenue, and backlog costs for each order, respectively. Then, we select a bar chart and define the data using the variables *C* (costs), *R* (revenue), and *Z* (backlog) as shown in ◘ Fig. 3.16.

In order to update the variables *R*, *C*, and *Z*, we should define some time points in the process flow diagram for this updating. For example, we can do it in the way shown in ◘ Fig. 3.17.

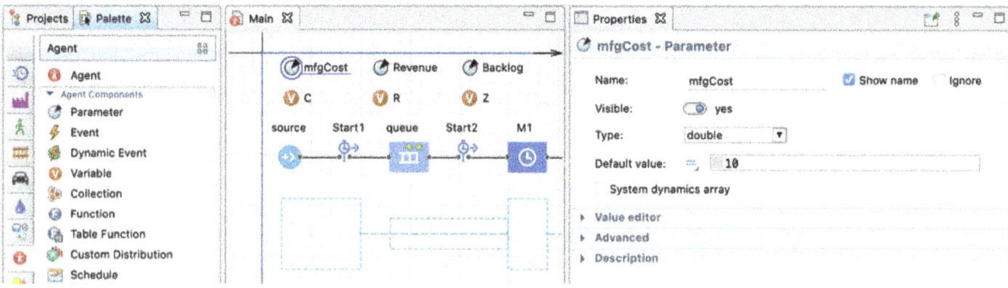


◘ **Fig. 3.15** Parameter and variable definition

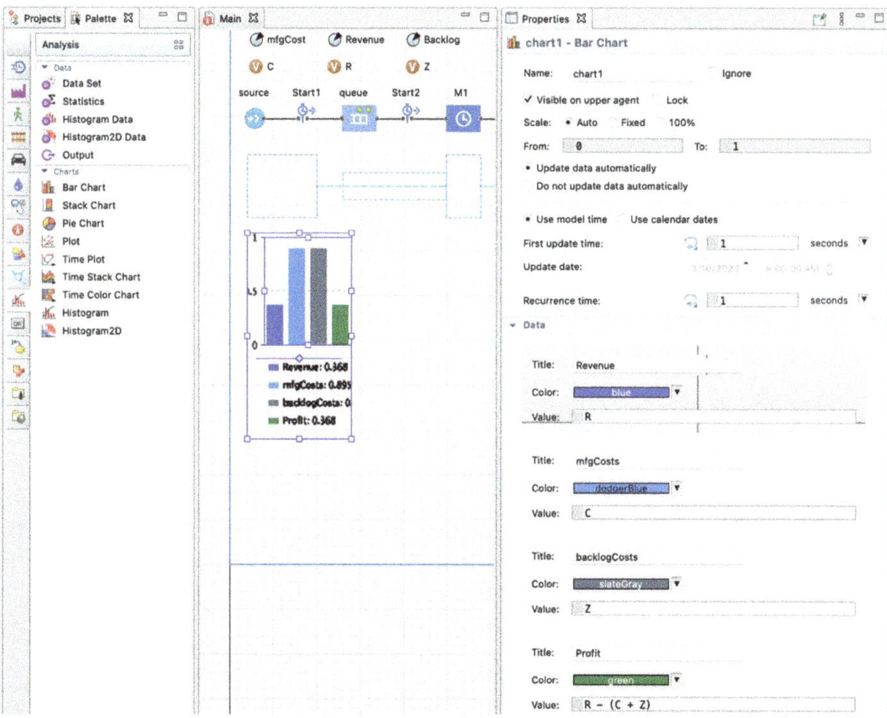


◘ **Fig. 3.16** Data definition for financial performance analysis

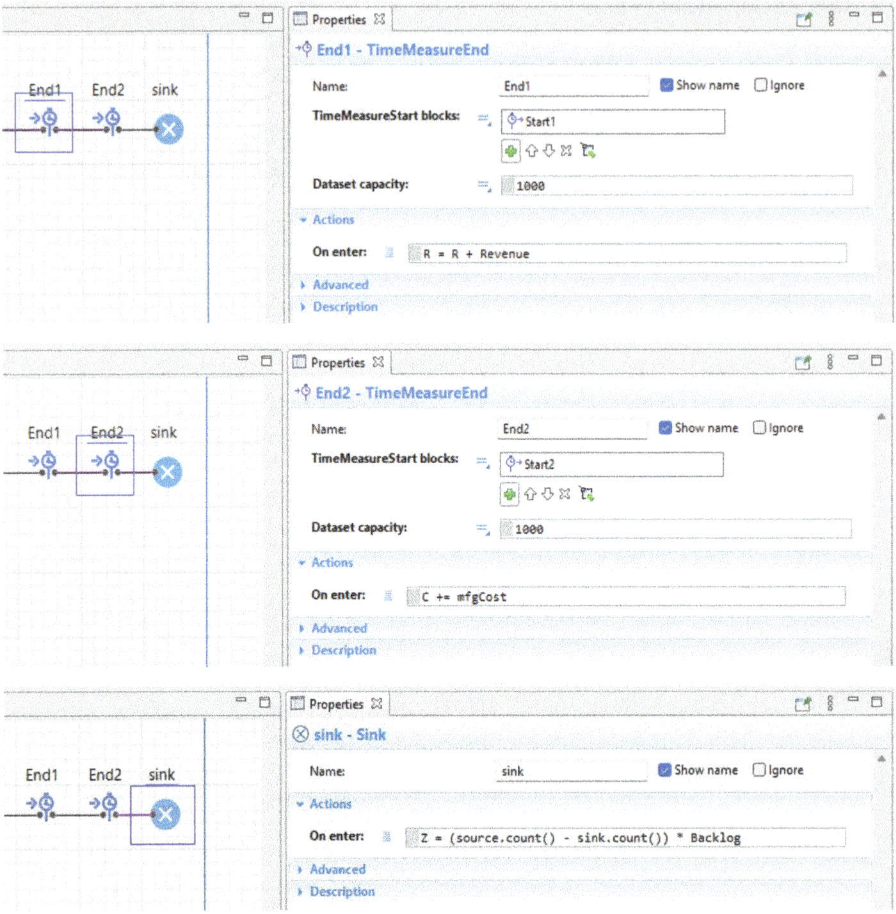


**Fig. 3.17** Variable update for financial performance analysis

There are more elegant ways to update variables, but the logic depicted in Fig. 3.17 is easy to understand. The AnyLogic functionality offers significantly broader options for defining and returning values, events, functions, and states using discrete event charts, agents, and system dynamics.

#### Step 2.2: Capacity Utilization

In the standard delay function, the statistics of average capacity utilization can be retrieved using the function *delay.statsUtilization.mean()*. We select the bar chart and create a diagram for capacity utilization analysis (Fig. 3.18).

In order to analyze the capacity utilization in dynamics, we create a time plot as shown in Fig. 3.19.

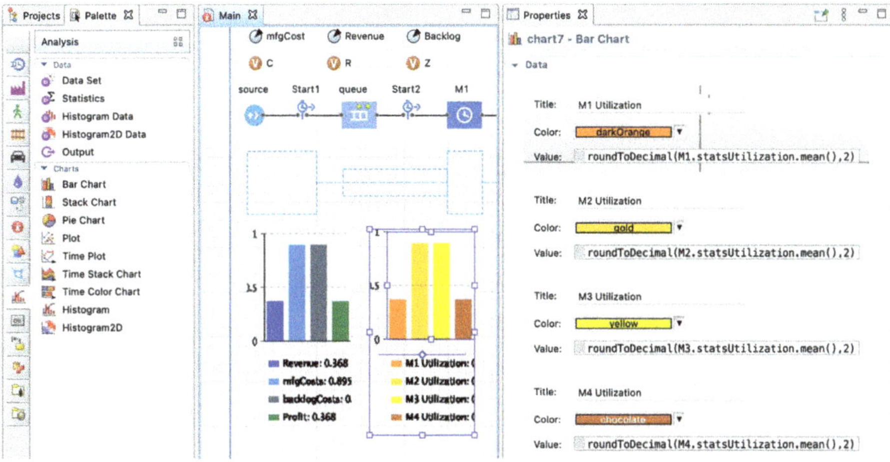


**Fig. 3.18** Data definition for capacity utilization analysis

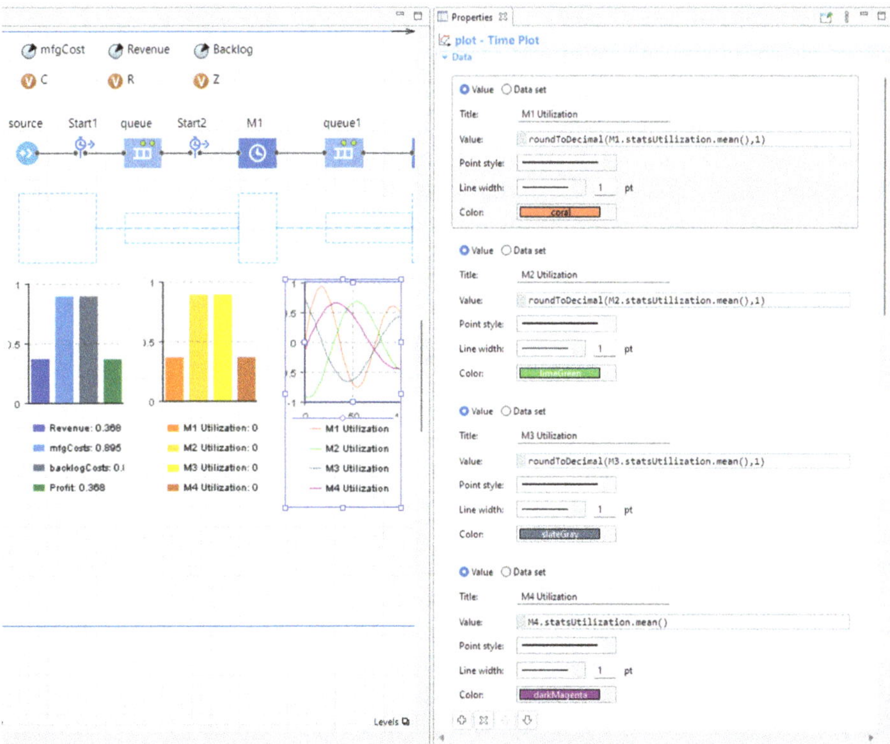


**Fig. 3.19** Value definition for capacity utilization analysis in time

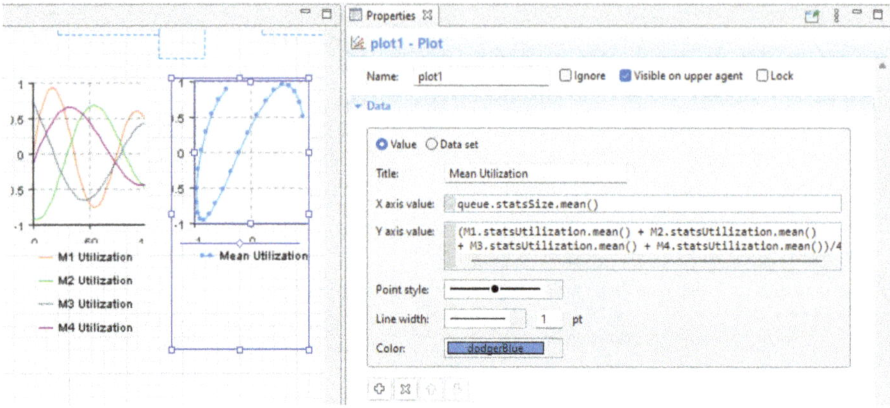


**Fig. 3.20** Value definition for mean capacity utilization analysis in dependence on average queue length in time

In order to estimate the dependence of the capacity utilization on the number of the waiting orders, we create a plot as shown in **Fig. 3.20**. In the standard queue function, the statistics of average queue length can be retrieved using the function *queue.statsSize.mean( )*.

In the standard Queue block, the statistics of the average queue length can be retrieved using the function Name.statsSize.mean(), where "Name" is the name of the Queue block. Thus, in this Plot, we set the *X*-axis value to queue.statsSize.mean() and the *Y*-axis value to the average utilization defined as shown in Eq. (3.16):

$$\begin{pmatrix} \text{M1.statsUtilization.mean}(\ ) + \text{M2.statsUtilization.mean}(\ ) \\ + \text{M3.statsUtilization.mean}(\ ) + \text{M4.statsUtilization.mean}(\ ) \end{pmatrix} / 4 \tag{3.16}$$

#### Step 2.3: Lead Time and Flow Time

In the process flow diagram, we included four points to measure the times. Recall that the point "start1" measures the time of customer order arrival, point "Start2" measures the time of customer order dispatching at the manufacturing line, and points "End1" and "End2" measure the time of order completion. The time period "End1 – Start1" is called *lead time*, and time period "End2 – Start2" is called *flow time. Lead time* roughly equals *flow time* + *waiting time.* If an order passes the End1 point, its lead time is written in the End1 dataset. Similarly, if an order passes the End2 point, its flow time is written in the End2 dataset. So, the task of collecting statistics on lead and flow times is now reduced to representing End1 and End2 datasets in the diagrams.

We select a time plot and create the flow time diagram as shown in **Fig. 3.21**.

Similarly, the lead time analysis diagram is created for the End1 dataset. To analyze the flow time distribution, a histogram can be created as shown in **Fig. 3.22**.

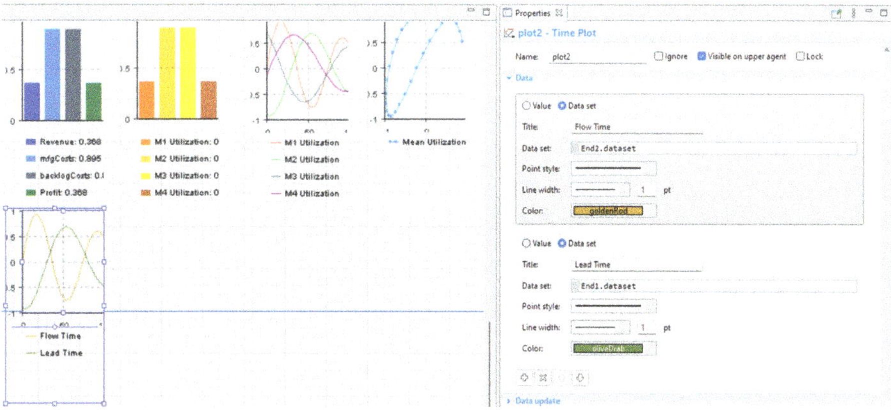


**Fig. 3.21** Flow time and lead time analysis

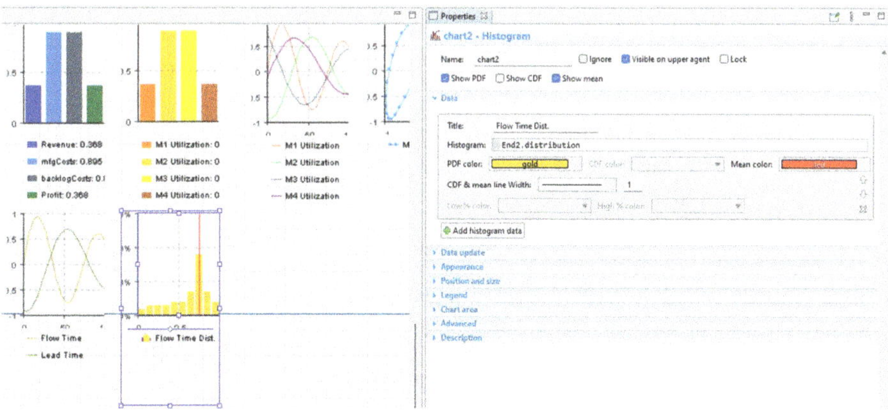


**Fig. 3.22** Flow time distribution analysis in time

Similarly, lead time distribution analysis diagram can be created for End1 data-set.

### Step 2.4: Backlog and Queue Analysis

First, we can analyze waiting orders and times. For analysis of waiting orders, we create a bar chart (Fig. 3.23).

Of course, we can also create a time plot (Fig. 3.24).

To compute the backlog, we use the function (*source.count( ) – sink.count( )*) (Fig. 3.25).

Now we created the KPI cockpit (Fig. 3.26).

*Note*: To enable statistics representation in the experiment view, the option "Force statistics collection" in each respective block of the process flow diagram should be enabled.

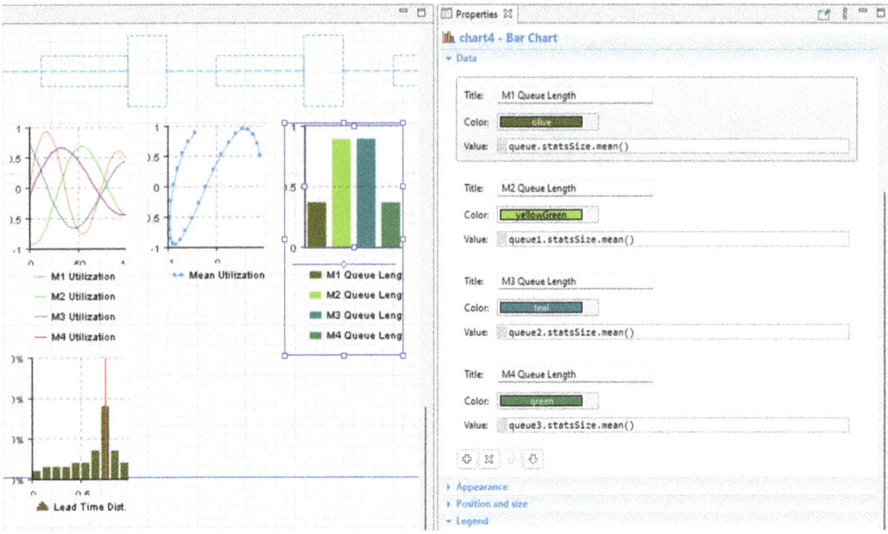


**Fig. 3.23** Waiting order analysis

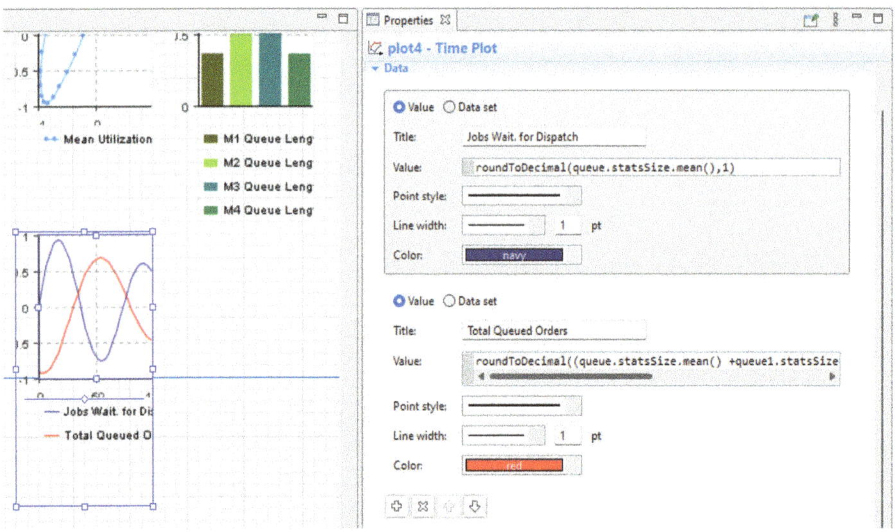


**Fig. 3.24** Waiting order analysis in time

## ■ Step 3: Creating an Interactive KPI Dashboard

So far, we changed the input data directly in the respective blocks of the model. Now we are going to create interfaces for playing the game more conveniently. In addition, we are going to learn some new methods for gathering statistics, more accurate and extensive cost analysis, and KPI computation.

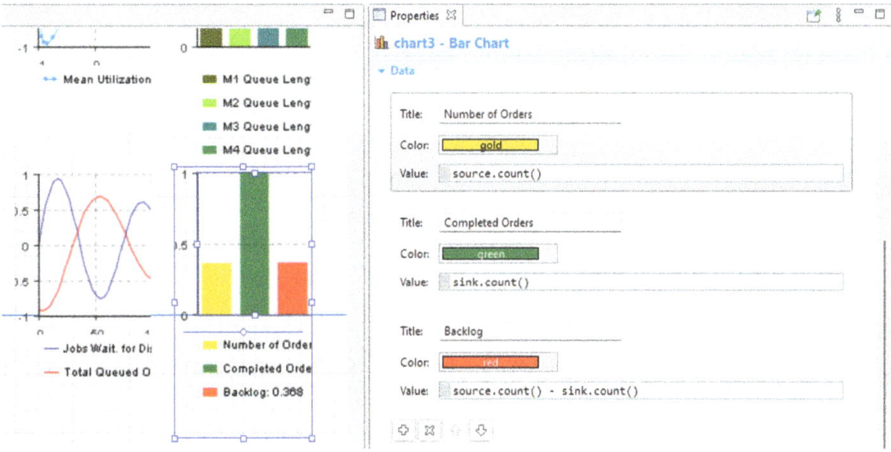


**Fig. 3.25** Backlog analysis

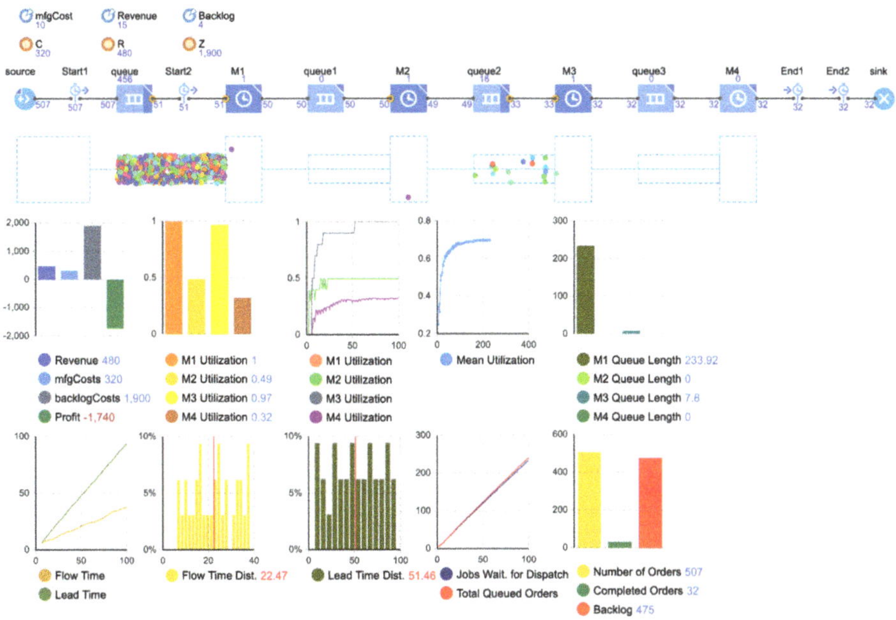


**Fig. 3.26** KPI dashboard for process dynamics analysis and decision-making

## Step 3.1: Views

You already might notice that we considered two large areas: process logic and output results in the form of statistics and diagrams for performance analysis. In many cases, it might be convenient to separate these views to reduce presentation and result interpretation complexity. This can be easily done. In "Presentation," we select "View" and place the pointer in the desired working area. In our example, we create two views: "Animation & Output" and "Process Logic" (◘ Fig. 3.27).

To define the actions for switching between the views, we should define the function "ViewName.navigateTo();" in the field "on click" of the respective text field. In the view "Process Logic," we will present the process flow diagram, define variable, parameters, function, events, and datasets and create control charts for real-time parameter changes during the simulation experiment later on. In the "Animation & Output view," the process visualization and KPI dashboard will be placed.



#### ▪▪ Step 3.2: Parameters

We start with the design of control charts for real-time parameter changes during the simulation experiment. For this purpose, we should first create new parameters (see ◘ Fig. 3.27) in addition to already created parameters "Costs," "Revenue," and "Backlog." We introduce four parameters for machine capacities (CapM1–CapM4), two parameters for minimal and maximal arrival quantity (MaxArrivalQuantity, ArrivalQuantity), arrival time (ArrivalTime), WIP inventory holding costs (InventoryCosts), opportunity costs (OpportunityCosts), exit on timeout duration (TimeOut), and processing times at the machines (MeanProcessTimeM1–MeanProcessTimeM2).

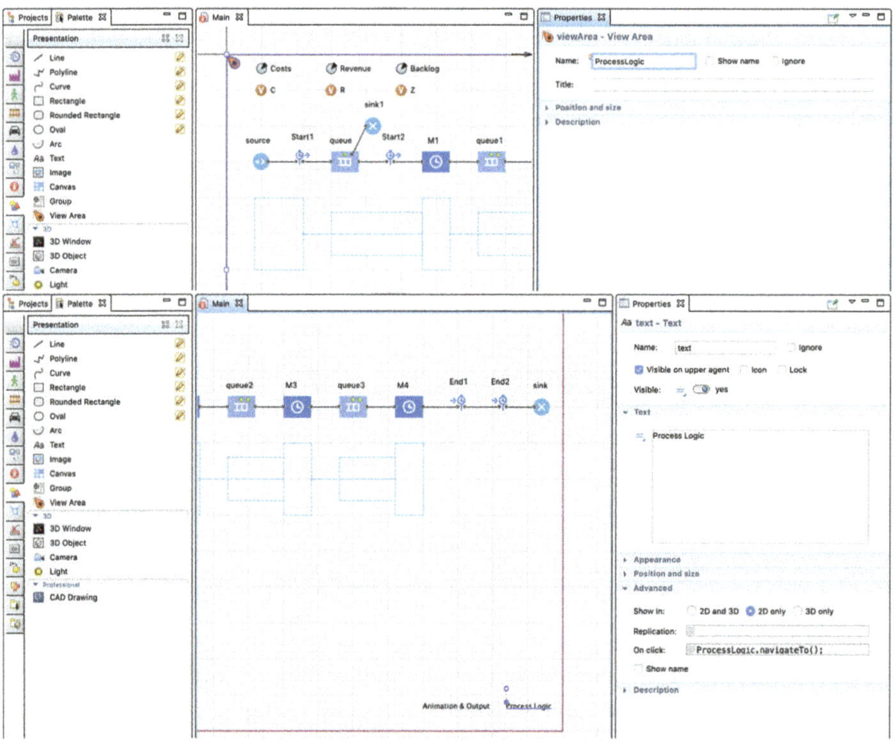


◘ **Fig. 3.27** Creation of different views in the model

#### ■■ Step 3.3: Variables

In the next step, variables should be defined in order to update the respective values during the experiment. We define the following variables:

- *C*—Manufacturing costs
- *R*—Revenue
- *Z*—Backlog costs
- *O*—Opportunity costs
- *I*—Inventory costs
- WIP—WIP inventory
- LO—Lost orders
- *Q*—Output quantity
- INPUT—Number of incoming orders
- Cap—Average capacity utilization

We define the places in the flow diagram where the variables will be updated, e.g., the sink node (◘ Fig. 3.28).

Variable LO is defined in the similar way in the "sink1," and the variable INPUT is defined in the element "Source."

#### ■■ Step 3.4: Functions

We now introduce some functions that are used to return values for statistics and KPI dashboard (◘ Fig. 3.29).

#### ■■ Step 3.5: Events

Finally, we would like to introduce a new method for statistics collection with the help of events. We need a diagram that would depict dynamics of incoming orders. For doing that, we need an event (named "event"), a dataset (named "Dataset"), and a variable (named "variable") (see ◘ Fig. 3.30). In ◘ Fig. 3.30, we define the rule for the event.

Since an event needs to be repeated (i.e., we want to know how many orders came each day or each week), we define this event as a "Cyclic" event. Setting a value in "Recurrence time" field, we define the cycle duration for updating the dataset. Now we can easily create a time plot for graphical presentation of incoming order dynamics (◘ Fig. 3.31).

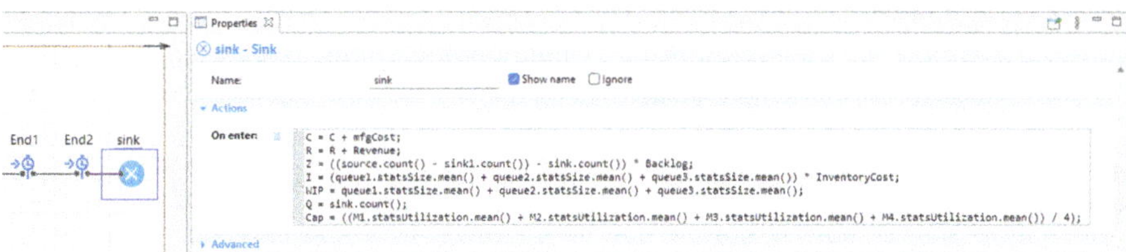


◘ **Fig. 3.28** Variable update definition

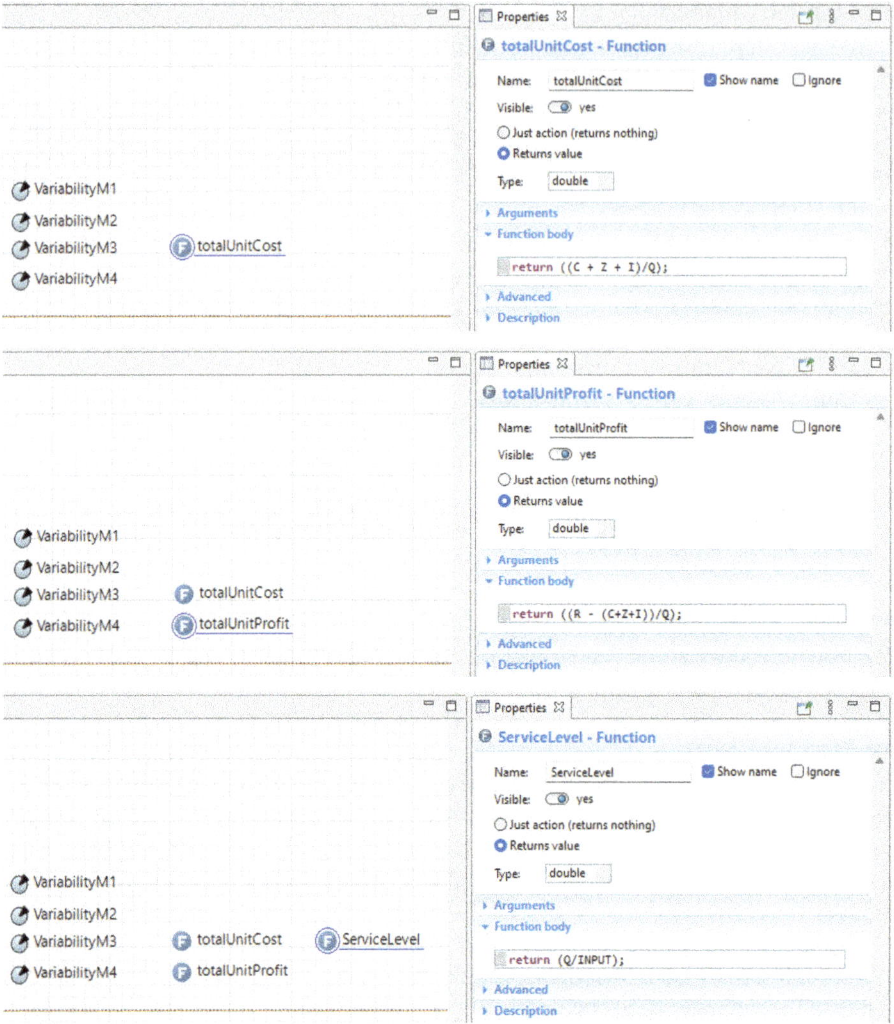


**Fig. 3.29** Function definition

### Step 3.6: Sliders

Sliders are useful for manual parameter change during the experiment. We define the sliders for all the parameters (Fig. 3.32).

In the properties, we should link the sliders to the respective parameter, define its min and max values, and press "Add the labels" to add "min," "value," and "max" texts on the slider.

### Step 3.7: KPI Dashboard Creation

In the "Animation & Output" view, we now design the KPI dashboard. We use the already created diagrams and extend the presentation by some KPIs (Fig. 3.33).

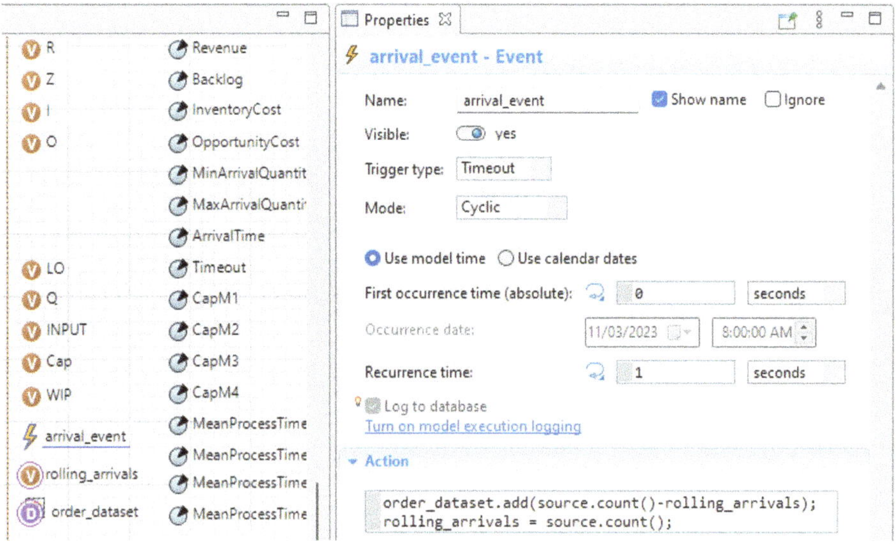


**Fig. 3.30** Event definition

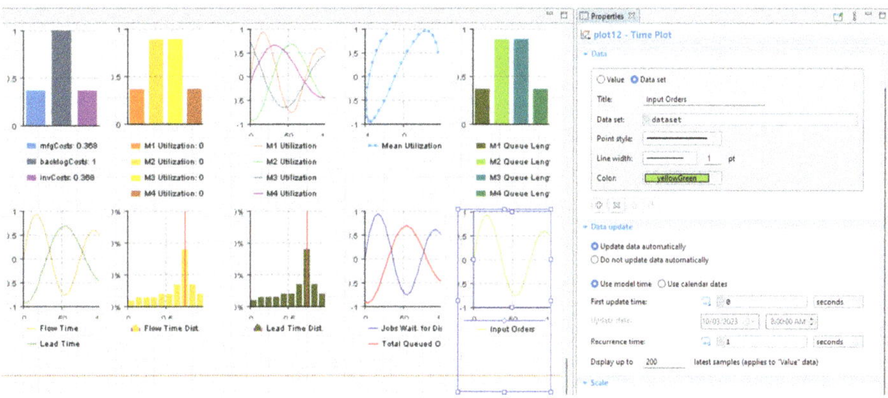


**Fig. 3.31** Input order dynamics evaluation

We widely use functions and variables for defining the values of the KPIs. An example for KPI "Unit Profit" is shown in Fig. 3.33.

### 3.2.4 Application Example

We now run our model (Model—Run). The simulation results over 100 s are shown in Fig. 3.34.

In the given setting, we achieved a total profit of $704, revenue of $1470, unit profit of $14, and unit cost of $16. We completed 49 orders, lost 73 orders, and

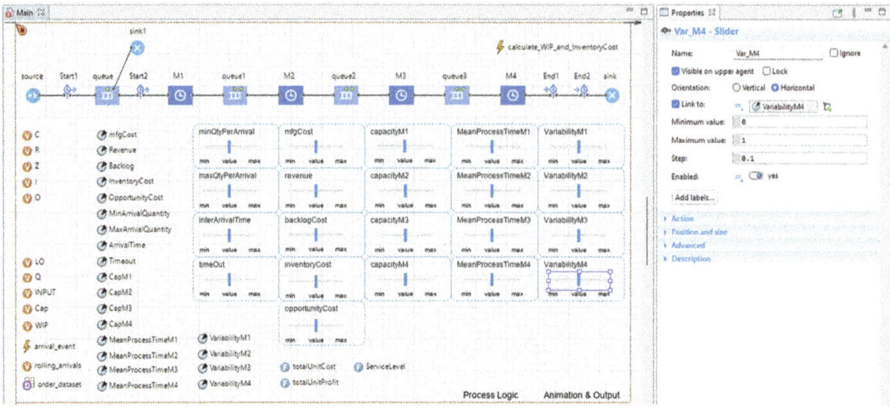


■ **Fig. 3.32** Slider definition

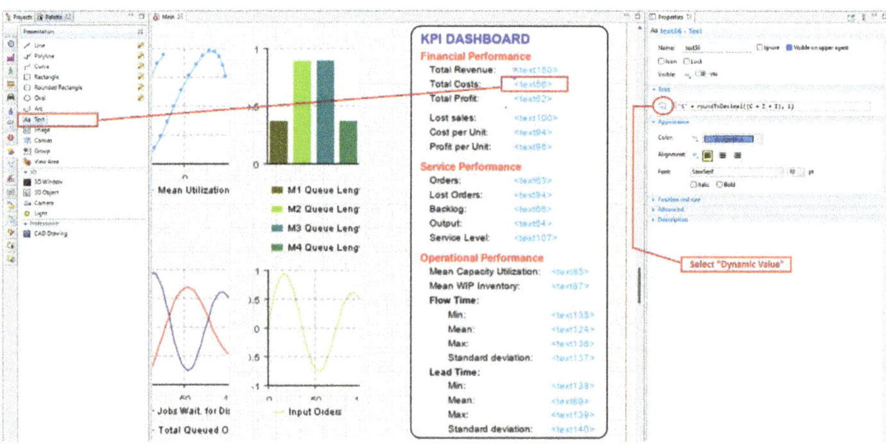


■ **Fig. 3.33** KPI dashboard

have a backlog of 46 orders. This implies a low service level of 29%. On the operational side, mean WIP inventory comprised 16.5 orders, mean capacity utilization is 7%, average flow time is 27 days, and average lead time is 37 days. We can observe from the visualization that the orders are backing up at M1 and M3 (see the top of Fig. 3.34 and the queue length KPI). There are two reasons for this. First, the incoming flow (1.75 orders per day) exceeds the capacity of M1 (1 order per day), on average. Second, the expected processing time at M3 is twice as long as the other machines; M3 is therefore *the process bottleneck*.

The results of round #1 and their analysis indicate the necessity to reduce the number of incoming orders and to increase the capacity of M3. We now test two options in the following setting: in the period between [30 and 55] s, the maximal arrival quantity will be reduced from 5 to 2 orders, and in the period from [35 to 60]

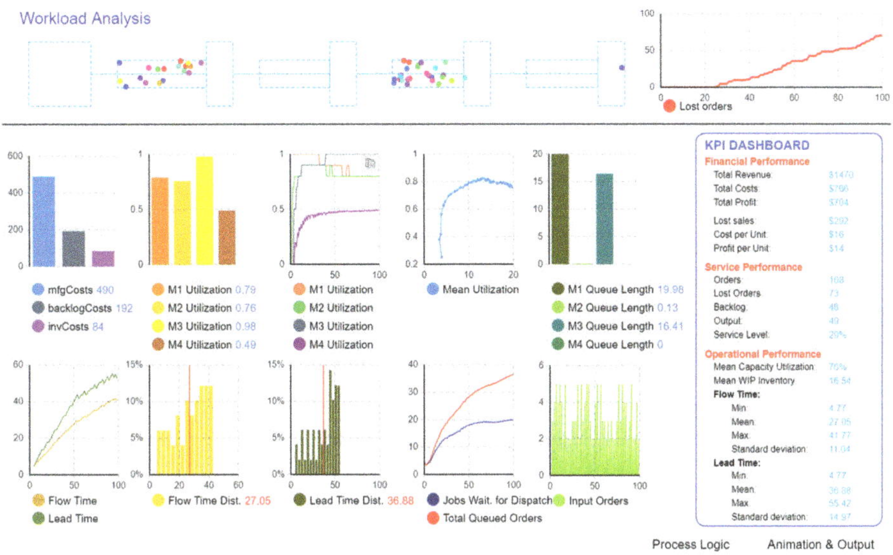


**Fig. 3.34** Results of round #1

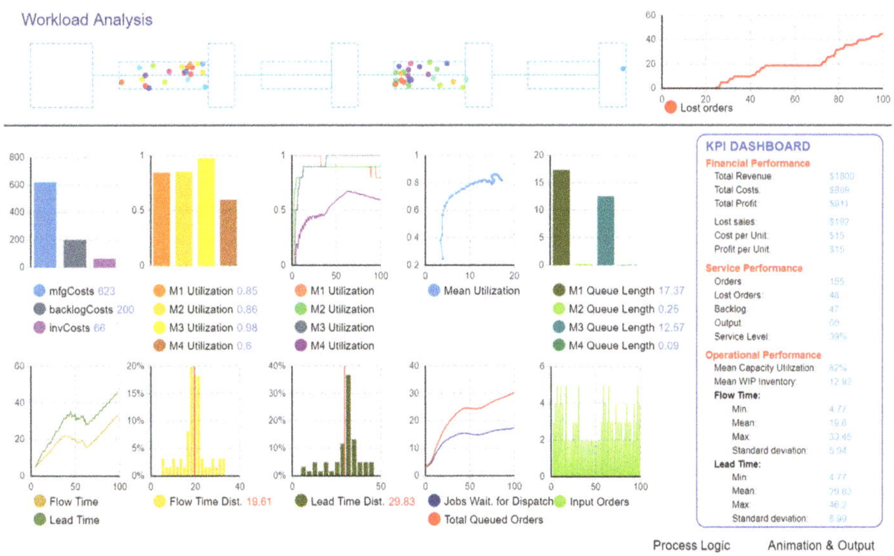


**Fig. 3.35** Results of round #2

s, the M3 capacity will be increased from 1 to 2 units, which implies manufacturing costs of \$11 for this period. We should use Sliders to change the parameters during the experiment at the respective intervals [30–55] and [35–60] s. The simulation results are presented in ◘ Fig. 3.35.

It can be observed that these changes improve the simulated service level (39% instead of 29%) and reduce lost orders (48 down from 73). We also improved the output (60 orders instead of 49), total profit ($911, up from $704), unit profit ($15 instead of $14), and unit costs ($15, not $16). At the same time, mean capacity utilization increased to 82%, average WIP inventory shrunk from 16.4 to 12.9 orders, average lead time reduced from 37 to 30 days, and average flow time from 27 to 20 days.

In the diagram "Lost orders," we can observe that the reduction of maximum arrival quantity in the period between [30 and 55] s positively affected the number of the lost orders. In the period between [45 and 65] s, the function is flat that indicates that no order was lost. Here, we can also observe a time delay (lag) between making a decision on parameter change and the performance impact of this change. In the diagrams "Flow time" and "Lead time," we also observe positive influence of the decisions to change the input rate and the M3 capacity.

Finally, we run a new experiment using a steady arrival quantity of 2 orders and new M3 capacity (two) with arrivals each 2 days. The simulation results are presented in ◘ Fig. 3.36.

As expected, we could further improve all the KPIs (◘ Table 3.2).

However, further improvements are possible! Make some suggestions on further process control improvements, and observe their performance impact through simulations.

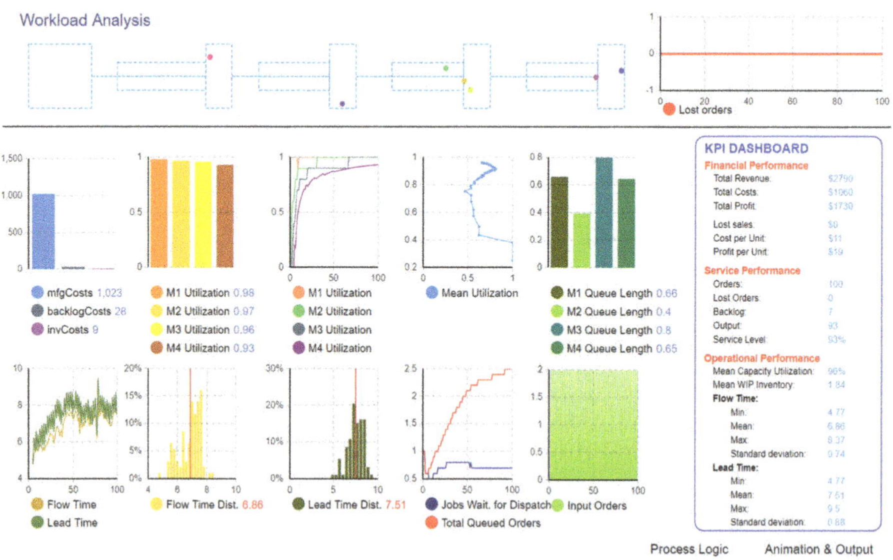


◘ **Fig. 3.36** Results of round #3

**Table 3.2** KPIs for financial and operational performance

| KPI | Round 1 | Round 2 | Round 3 |
|---|---|---|---|
| Total profit, $ | 704 | 911 | 1730 |
| Unit profit, $ | 14 | 15 | 19 |
| Unit costs, $ | 16 | 15 | 11 |
| Flow time, days | 27.1 | 19.6 | 6.9 |
| Average WIP inventory | 16.5 | 12.9 | 1.8 |
| Total sales, $ | 1470 | 1800 | 2790 |
| Output, units | 49 | 60 | 93 |
| Lost orders, units | 73 | 48 | 0 |
| Lead time, days | 36.9 | 29.8 | 7.5 |
| Service level, % | 29 | 39 | 93 |
| Average capacity utilization, % | 76 | 82 | 96 |
| Backlog, units | 46 | 47 | 7 |

## 3.3 Case Study "EOQ and Inventory Control" in AnyLogic

### 3.3.1 Development of an EOQ Model in AnyLogic

We start by analyzing a special case of ($s$, $q$) models when $q$ is optimized by the economic order quantity (EOQ) model (see ▶ Chap. 2). We create a simulation model for the EOQ problem using events; the dynamics of daily inventory consumption and inventory replenishment are modeled using an AnyLogic event named Ordering. We also add to the model one parameter OrderQuantity, and two variables called "Demand" and "InventoryLevel," and a time plot of inventory levels (◘ Fig. 3.37).

In event "Ordering," we describe the logic of EOQ model (◘ Fig. 3.38).

The event "Ordering" is a cyclic event with recurrence time of 1 s (e.g., one day of real life). According to the EOQ model, inventory consumption is a linear function from demand and that is why we describe inventory dynamics as:

```
InventoryLevel -= Demand;
```

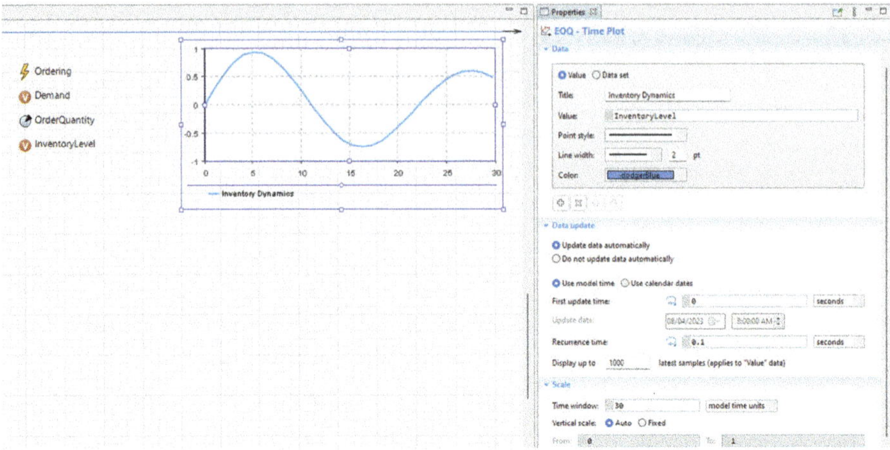


**Fig. 3.37** EOQ model

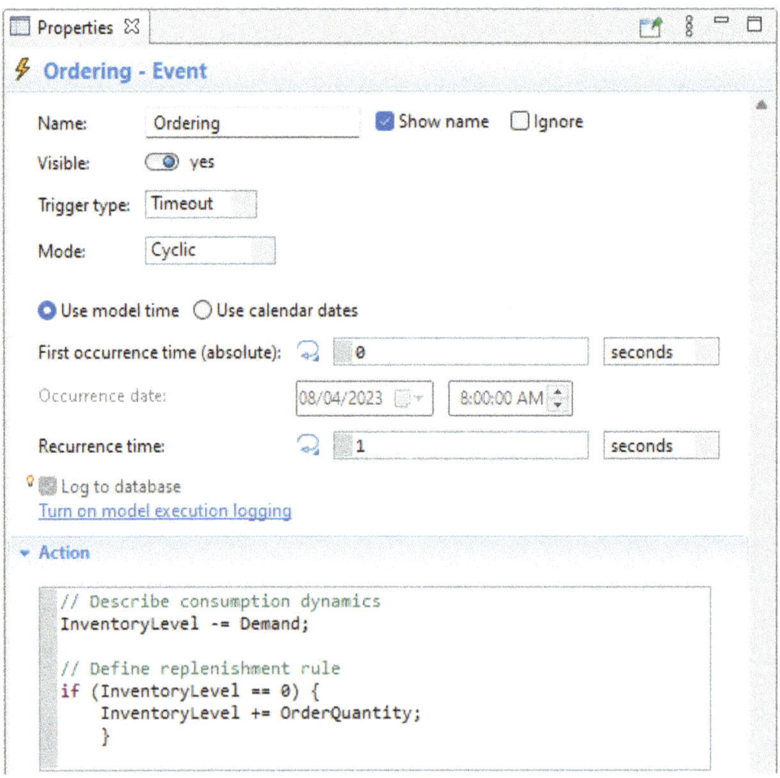


**Fig. 3.38** Event description

According to EOQ, if inventory level is zero, new order of the same quantity (*q*) arrives immediately. That is why we describe replenishment rule using Java as follows:

```
if (InventoryLevel == 0) {
InventoryLevel += OrderQuantity;
}
```

Setting "Demand" = 80 and "OrderQuantity" = 400, we run the experiment and observe inventory dynamics (◘ Fig. 3.39).

It can be observed from ◘ Fig. 3.39 that each day inventory level is reduced by demand. At the fifth day, inventory level is zero, and immediate replenishment is generated. That is why the step from 80 to 0 is missing in the diagram in ◘ Fig. 3.39.

### 3.3.2 Development of Periodic Inventory Control Model in AnyLogic

While the EOQ simulation example deals with *cycle inventory*, we now introduce *safety stock* into consideration (see ► Chap. 2). Assume that daily demand and/or lead time are normally distributed. We need to define additional parameters for lead time, mean demand, mean lead time, standard deviation of daily demand, and standard deviation of lead time. In "Ordering" event, we add (◘ Fig. 3.40):

```
Demand = round( max (normal(DevDemand,MeanDemand),0));
LeadTime = round( max (normal(DevLT,MeanLT),0));
```

First, consider the case without lead time. The simulation result for mean demand = 80 units and standard demand deviation = 18 is presented in ◘ Fig. 3.41.

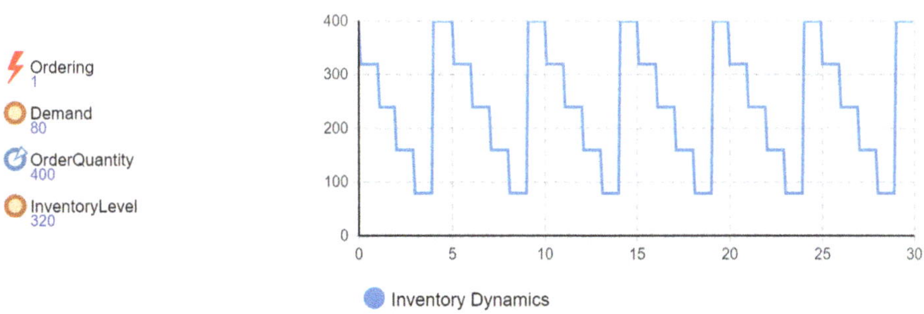


◘ **Fig. 3.39** EOQ inventory dynamics

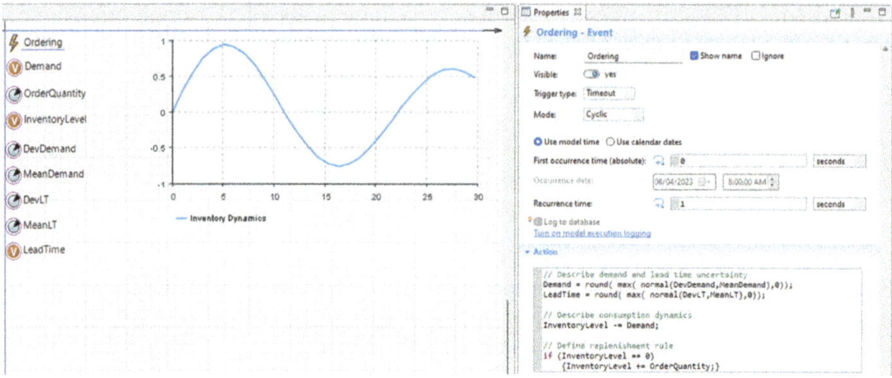


**Fig. 3.40** Stochastic inventory management simulation

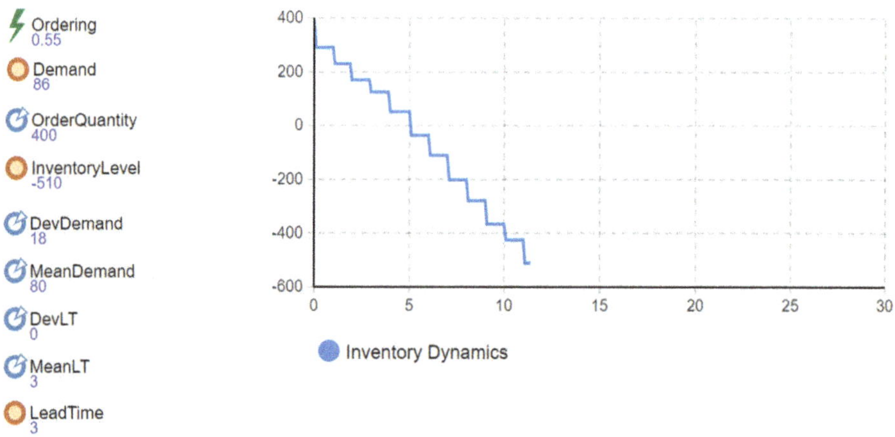


**Fig. 3.41** Stochastic inventory management simulation results

It can be observed from Fig. 3.41 that no replenishment actions are performed, and the model works wrong. The reason is the condition for new replenishment defined in "Ordering" event:

```
// Define replenishment rule
if (InventoryLevel == 0){
InventoryLevel += OrderQuantity;
}
```

Due to demand deviations, the condition InventoryLevel = 0 will not be fulfilled. That is why we have to modify this condition.

For *periodic review system*, this condition is not needed at all since we replenish at fixed dates. Thus, we just modify the recurrence time of the event "Ordering" by defining it through the parameter "t_Period." This makes the Ordering event recur at a time scale that is different than daily demand, and so we introduce a new event named Consumption dynamics to capture the daily demand consumption of inventory. ◘ Figure 3.42 shows the logic of the Consumption dynamics and Ordering events needed for the periodic review policy ($t$, $q$).

◘ Figure 3.42 describes periodic review policy $t$, $q$. For $t$, $S$ policy, the line

```
InventoryLevel += OrderQuantity;
```

should be re-written as

```
InventoryLevel = TargetInventory;
```

where TargetInventory is a new parameter to define $S$. The value of this parameter can be fixed by a number or defined dynamically by a condition (see further in this chapter).

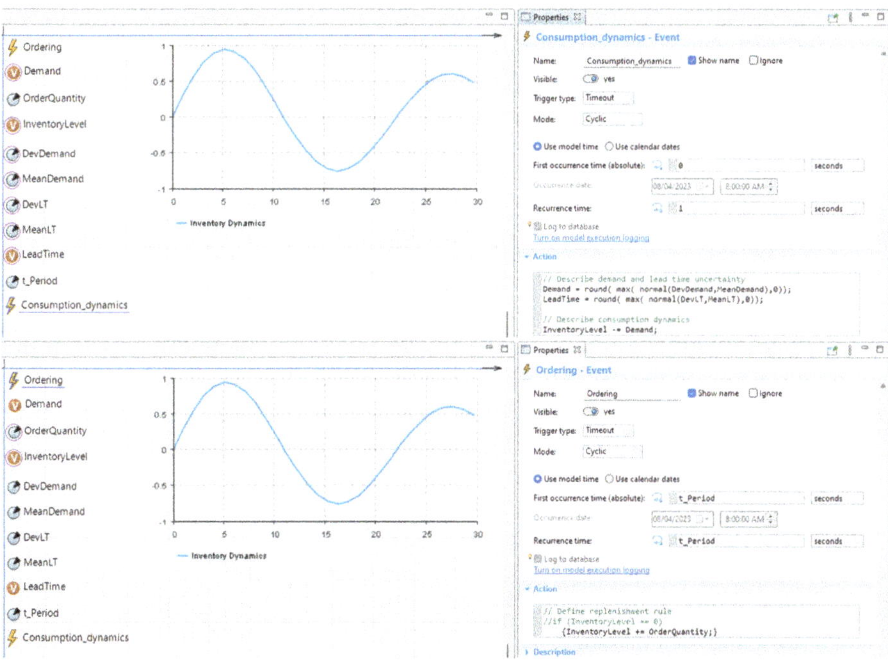


◘ **Fig. 3.42** Periodic review policy $t$, $q$

### 3.3.3 Development of a Continuous Inventory Control Model in AnyLogic

For *continuous review system*, replenishment rule can be defined with the help of reorder point (ROP). Thus, we modify the condition.

3

```
// Define replenishment rule
if (InventoryLevel == 0){
InventoryLevel += OrderQuantity;
}
```

as new condition using a new parameter "ROP":

```
// Define replenishment rule
if (InventoryLevel <= ROP){
InventoryLevel += OrderQuantity;
}
```

As such, *s*, *q* policy is defined. The simulation results for ROP = 120 units, mean demand = 80 units, and standard deviation of daily demand = 18 are presented in ◘ Fig. 3.43.

In order to define the *s*, *S* policy, the line.

```
InventoryLevel += OrderQuantity;
```

needs to be rewritten as

```
InventoryLevel = TargetInventory;
```

where TargetInventory is the parameter to define *S*. The value of this parameter can be fixed or defined dynamically by an ROP-dependent condition subject to changes in demand and lead time (see further in this chapter).

The real meaning of ROP becomes obvious if we extend our models by lead time. In addition, we will include safety stock in our model.

For modeling lead time, we define new event "NewOrder" and new variable "OrderInTransit" of type "Boolean" (with initial value "false") and rewrite replenishment rule in the event "Ordering" as follows (◘ Fig. 3.44):

```
// Define replenishment rule
if (InventoryLevel <= ROP && !OrderInTransit)
{NewOrder.restart( LeadTime );
    OrderReceived = true;
}
```

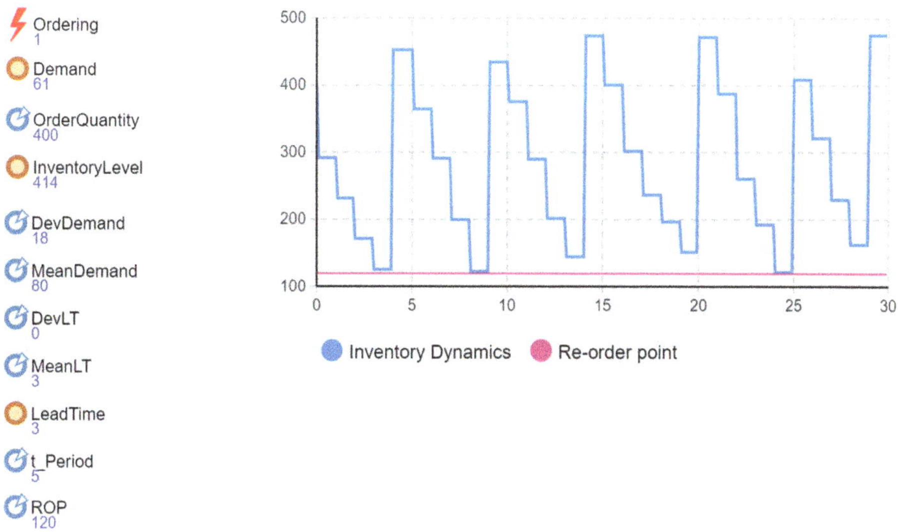


**Fig. 3.43** Continuous review policy *s*, *q*

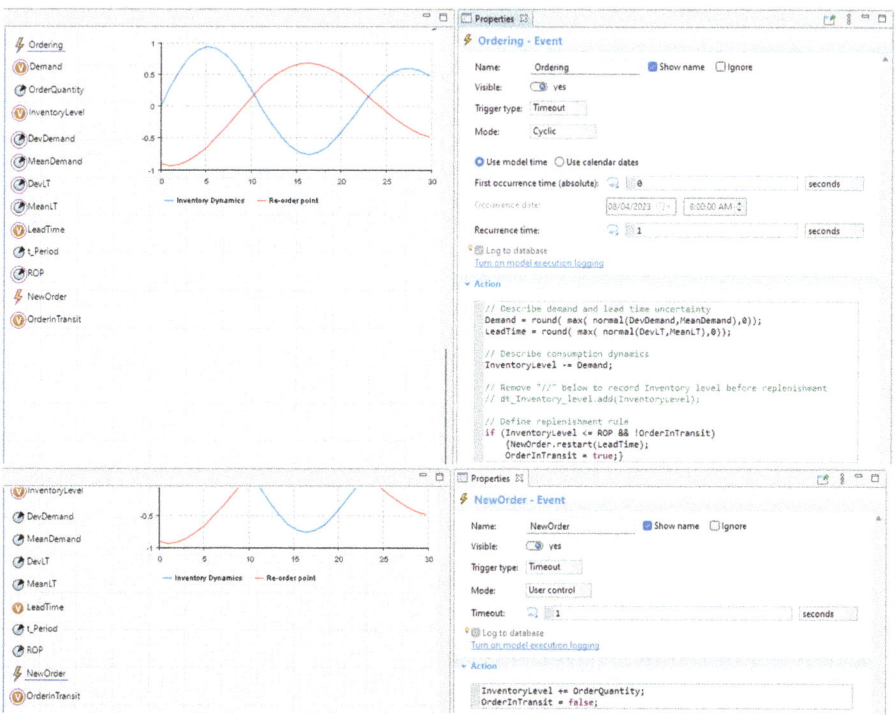


**Fig. 3.44** Continuous review policy *s*, *q* with lead time

We define an event called "NewOrder" with the action logic depicted in ◘ Fig. 3.44 given by

```
InventoryLevel += OrderQuantity;
OrderInTransit = false;
```

3

Here, we used two new Java operators:

```
«&&» - logical "and"
«!» - logical "not"
```

If inventory level reaches the ROP, new replenishment order is generated, and new order arrives in *n*-days defined in the LeadTime parameter. That is why we let the event "NewOrder" start in *n*-days defined in the LeadTime parameter. At the time of order entry and inventory level increase, the logical variable "OrderInTransit" gets the value "false." This means that no other orders are in transit. Next order will be generated when inventory level reaches the ROP again. We need the component "OrderInTransit" in the condition **if** (InventoryLevel<=ROP && !OrderInTransit), since new order is generated if both ROP is reached and there are no other orders in transit.

The simulation result for ROP = 240 units and mean LT = 3 days (without deviation) is presented in ◘ Fig. 3.45.

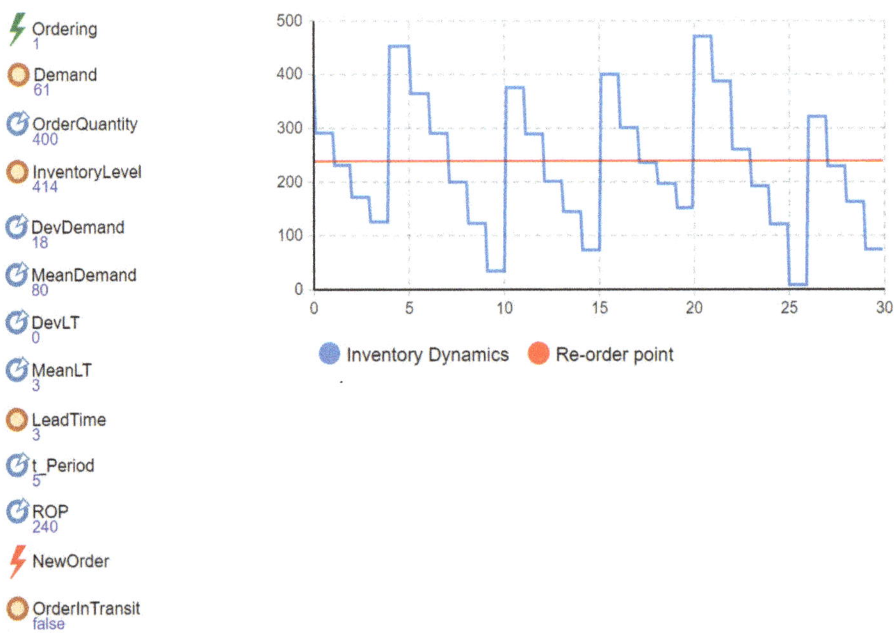


◘ **Fig. 3.45** Simulation of continuous review policy *s*, *q* with lead time

In order to estimate the efficiency of different ordering policies, we should estimate inventory holding, ordering, and stock-out costs. First, new parameters and variables for these three types of costs are needed. Second, diagrams should be constructed as bar charts and time plots.

We define three new parameters "HoldingCost," "OrderingCost," and "StockoutCosts" and three new variables "HoldingCosts," "OrderingCosts," and "StockoutCosts." Further, three new variables "FillRate," "TotalSales," and "RetailShortage" are defined.

In event "Ordering," we add a new condition:

```
// Compute sales and retail shortage
if (InventoryLevel >= Demand)
   {TotalSales += Demand;}
else
   if (InventoryLevel >= 0)
      {TotalSales += InventoryLevel;
      RetailShortage += Demand - InventoryLevel;}
   else
      {RetailShortage += Demand;}
```

Then, we create a new event "CostUpdate" as shown in ◘ Fig. 3.46.

Costs, shortage, and fill rate are updated in the event "CostsUpdate." Note that holding costs are due to on-hand inventory, if any, at the end of each day. We assume that there are no lost sales and that all demand is backlogged during shortages.

Finally, we create two time plots for variables "FillRate" and "RetailShortage" and a stack chart for holding, ordering, and stock-out costs.

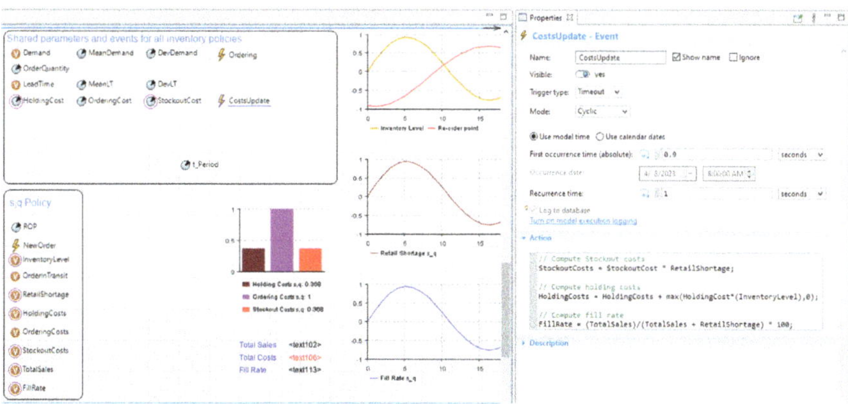


◘ **Fig. 3.46** Costs, shortage, and fill rate update event

3

The simulation results for ROP = 120 units, holding costs = \$0.10 per day per unit, stock-out costs = \$1 per unit, and ordering costs = \$90 per order are presented in ◘ Fig. 3.47.

A simulation experiment of these parameters for 30 s and random seed 1 is presented in ◘ Fig. 3.47. It can be observed that total costs are \$1743; the average fill rate decreases during the experiment, 56.7% by month's end; and the cumulative retail shortage reaches 1005 unit-days by the end of the month.

Let us enhance the analysis from ◘ Fig. 3.47 by computing optimal ROP with the use of safety stock (see ► Chap. 2). For example, 0.05 probability of stock-out corresponds to a 95% service level.

Let us compute ROP subject to service level 95% using Eq. (► 2.25, see ► Chap. 2):

$ROP = 80 \times 3 + 1.65 \times 18 \times \sqrt{3} = 240 + 52 = 292$ units, where 52 units is safety stock. For computing order quantity, we use EOQ formula:

$$q = \sqrt{\frac{2 \times 2400 \times 90}{3}} = 379 \text{ units}$$

The simulation results with new $s = 292$ units and $q = 379$ units are presented in ◘ Fig. 3.48.

It can be observed from ◘ Fig. 3.48 that total cost equals \$1405 (as compared to \$1743 in the initial situation), fill rate increased from 56.7% to 95.5% at the end of the month, and the cumulative retail shortage decreases from 1005 to 105 units.

If managers desire further stock-out protection, they may consider raising the reorder point to achieve, say, 99% service level. Substituting $z = 2.33$ yields:

$$ROP = 80 \times 3 + 2.33 \times 18 \times \sqrt{3} = 240 + 73 = 313 \text{ units}$$

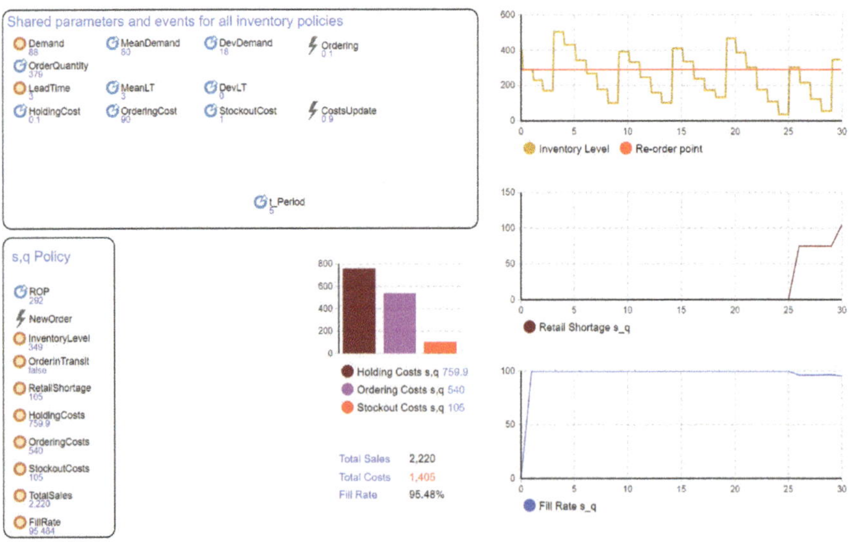


◘ **Fig. 3.47** Efficiency analysis for *s*, *q* policy

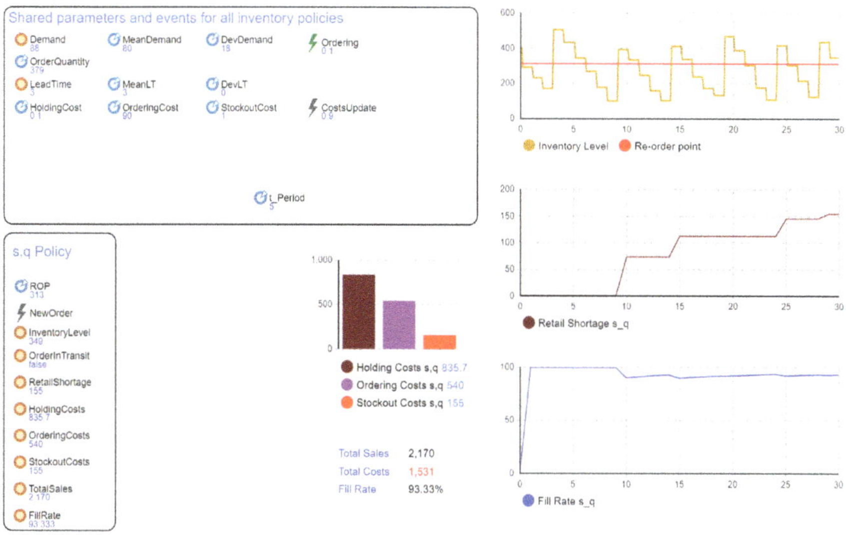


■ **Fig. 3.48** Efficiency analysis for optimized *s*, *q* policy

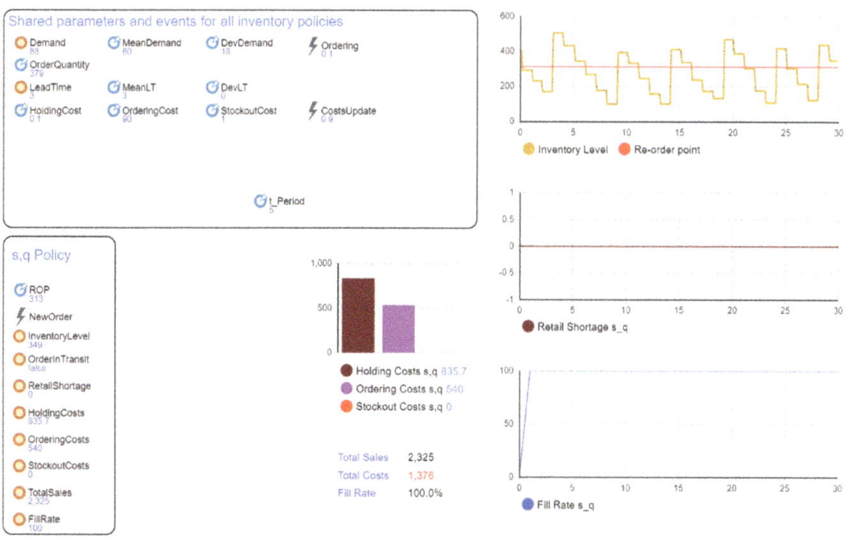


■ **Fig. 3.49** Efficiency analysis for optimized *s*, *q* policy

The results of rerunning the simulation with order quantity $q = 379$ and reorder point ROP = 313 are presented in ■ Fig. 3.49; it can be observed that total costs dropped by 2% to \$1376 and average fill rates increased from 95.5% to 100% (stockouts were eliminated).

3

### 3.3.4 Events and Application Example

In order to decouple inventory replenishment logic for different control policies, we introduce action charts for continuous review policies instead of placing all the codes in event "Ordering."

Action charts are created in "Palette → Actioncharts" (◘ Fig. 3.50).

In action chart, we can define a condition for activating the action chart and the code (action) to be performed if the condition is true or false. In our case, we define a condition "Order placed" as

```
InventoryLevel <= ROP && !OrderInTransit
```

and the code to be executed for "Update Inventory" if the condition "Order placed" is true as

```
NewOrder.restart( LeadTime );
OrderInTransit = true;
```

Previously, we had this definition of replenishment rule directly in the event "Ordering" (see ◘ Figs. 3.38 and 3.44):

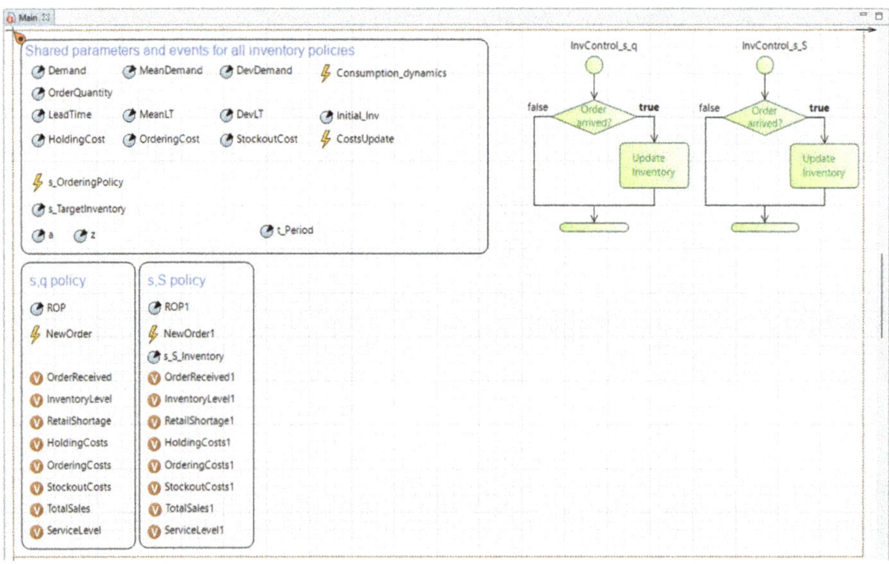


◘ **Fig. 3.50** Action chart for *s*, *q* policy and *s*, *S* policy

```
// Define replenishment rule
if (InventoryLevel <= ROP && !OrderInTransit)
{NewOrder.restart( LeadTime );
    OrderInTransit = true;
}
```

Usage of action charts allows decentralizing different replenishment actions and therefore decoupling inventory dynamics in different control policies. Action charts may have much more complex structure with numerous conditions, branches, and actions.

For our case, we create two action charts for *s*, *q* and *s*, *S* policies calling the already existing event "NewOrder" for *s*, *q* policy and new event "NewOrder1" for *s*, *S* policy. The event "NewOrder1" is described as:

```
InventoryLevel1 = s_TargetInventory;
OrderingCosts1 = OrderingCosts1 + OrderingCost;
OrderReceived1 = false;
```

For *t*, *q* and *t*, *S* policies, no action charts are needed. We just create a new event with the following code:

```
// Count time
t2++;
t3++;

// Periodic Review - Fixed Quantity Model
if (t2 == t Period)
{InventoryLevel2 += OrderQuantity;
OrderingCosts2 = OrderingCosts2 + OrderingCost;
t2 = 0;}

// Periodic Review - Target Inventory Model
// Calculate t Target Inventory
t TargetInventory = MeanDemand*(t Period + MeanLT) +
                    z*DevDemand*sqrt(t Period + MeanLT);

// Simple logic with deterministic leadtime
if (t3 == t Period - LeadTime)
{Order quantity3 = (t TargetInventory - InventoryLevel3);}

if (t3 == t Period)
{InventoryLevel3 += Order quantity3;
OrderingCosts3 = OrderingCosts3 + OrderingCost;
t3 = 0;}
```

Previously, we introduced new parameters s_TargetInventory and t_TargetInventory in order to define target inventory. The target inventory can be fixed or dynamic. If we fix the value of target inventory, this value will be used over the whole experiment. From the management point of view, such a policy can be quite efficient in regard to fixed warehouse area and transportation planning. However, if demand changes over the considered standard deviation, shortages can occur. The same holds true for ROP.

That is why it can be useful to change both ROP and target inventory dynamically. One possible option is to establish a relation between ROP and target inventory. Since we have different inventory dynamics in *s*, *q* and *s*, *S* policies, we introduce an additional parameter ROP1 for *s*, *S* policy, while ROP will be used for *s*, *q* policy.

We also introduce a new parameter s_S_inventory that will keep the current inventory level (variable InventoryLevel1) at the time of new order launching.

Now we update the code "UpdateInventory" in action chart InvControl_s_S as follows:

```
S_TargetInventory = (ROP1-InventoryLevel1) + (a*(MeanDemand+Dev
Demand)*LeadTime);
NewOrder1.restart( LeadTime );
OrderInTransit1 = true;
```

It is noted that we set the default value of parameter $a = 1$ in the formula that computes s_TargetInventory.

The rule

```
s_TargetInventory = (ROP1 - InventoryLevel1) +
    (a*(MeanDemand + DevDemand)*LeadTime);
```

for setting target inventory in *s*, *S* policy is our subjective estimation. It can be adjusted during experiments. Note that we also added a new logical variable OrderReceived1 for *s*, *S* policy to decouple *s*, *q* and *s*, *S* policies.

The event "Consumption_dynamics" has now the following form (◘ Fig. 3.51).

For dynamic updates of ROPs, we use ► Eq. (2.25). We separate logic and output analysis views (◘ Figs. 3.52 and 3.53).

Let us specify data for application example. We are going to compare four inventory control policies for each product (see Chap. 2, Section 2.2.5). We will compare for all products:

- *t*, *q* policy with fixed period and quantity
- *t*, *S* policy with fixed period and target inventory
- *s*, *q* policy with dynamic ROP and fixed order quantity
- *s*, *S* policy with dynamic ROP and target inventory

In all experiments, demand will be assumed to be variable, and lead time will be assumed to be constant. z-value will be 1.65 (according to service level = 95%). Holding cost is $0.10 per unit per day, ordering cost is $90 per order, and stock-out cost is $0.20 per day per unit.

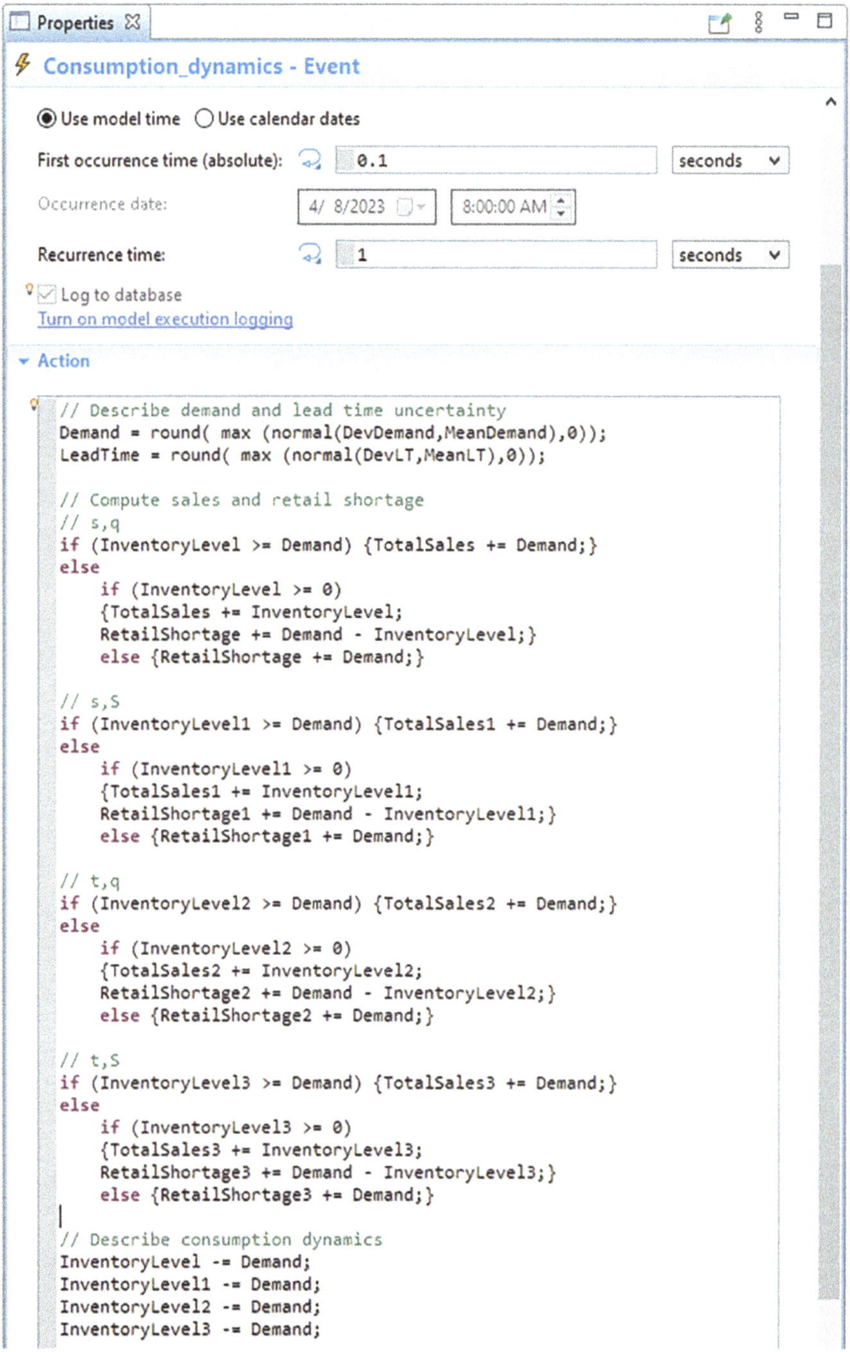


**Fig. 3.51** Event "Consumption_dynamics"

### Experiment 1: Impact of Demand Dynamics

The following input data is used (Table 3.3).

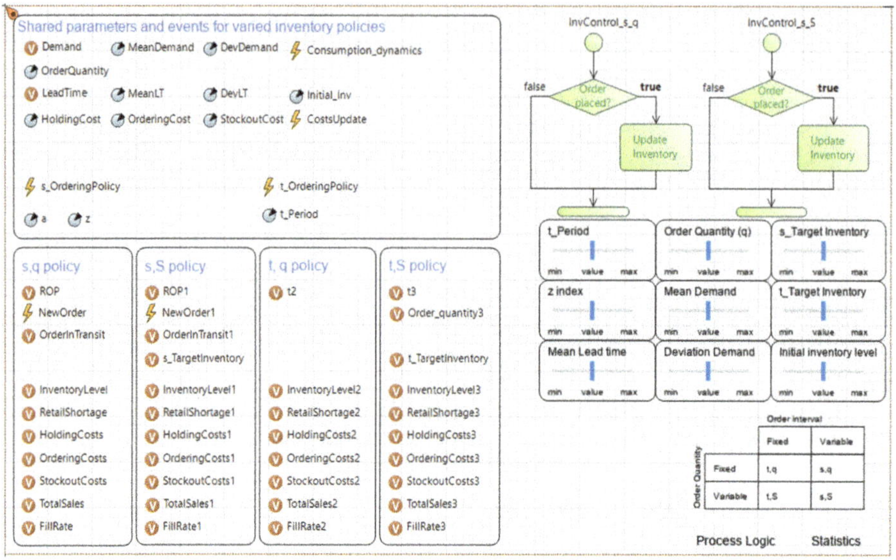


■ **Fig. 3.52** Logic view

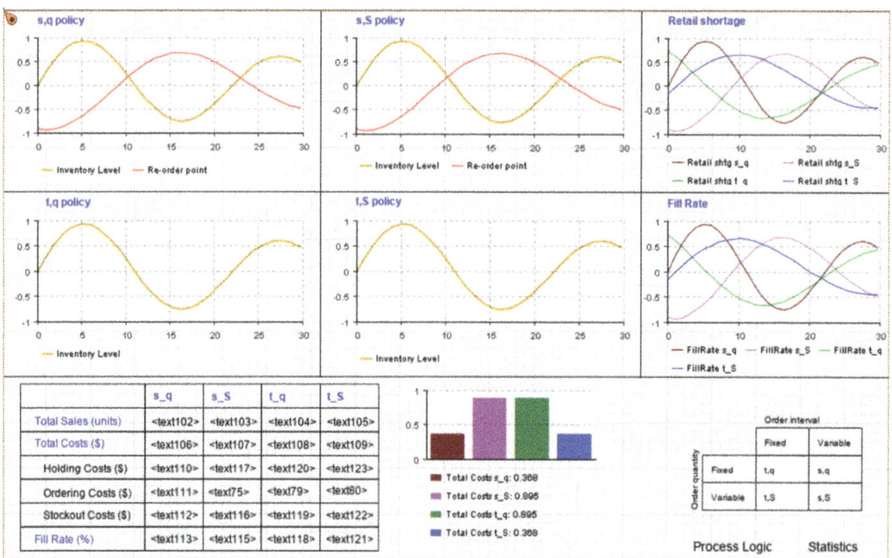


■ **Fig. 3.53** Analysis view

The simulation results are presented in ■ Figs. 3.54 and 3.55.

It can be observed from ■ Figs. 3.54 and 3.55 that *s*, *S* is the most efficient policy with a cost of $1144, while periodic review policy *t*, *q* provides 100% service level but at the same time has the highest cost of $3109.

**Table 3.3** Input data for experiment 1

| Parameter | Description | Value |
|---|---|---|
| OrderQuantity | Order quantity, units | 1000 |
| MeanDemand | Mean daily demand, units | 200 |
| DevDemand | Std. dev. of daily demand, units | 40 |
| MeanLT | Lead time, days | 3 |
| DevLT | Std. dev. of lead time | 0 |
| *t* | Time between orders, days | 5 |
| HoldingCost | Holding cost per unit per day | 0.10 |
| OrderingCost | Fixed ordering cost | 90 |
| StockoutCost | Backlogging cost per unit per day | 0.20 |
| t_TargetInventory | Order-up-to inventory target, units | Dynamic |
| s_TargetInventory | Order-up-to inventory target, units | Dynamic |
| *z* | *z*-value for 95% service level | 1.65 |
| Initial | Inv initial inventory level, units | 1200 |
| *a* | Parameter in s_TargetInventory rule | 1 |
| ROP | Reorder point, units | Dynamic |
| ROP1 | Reorder point, units | Dynamic |

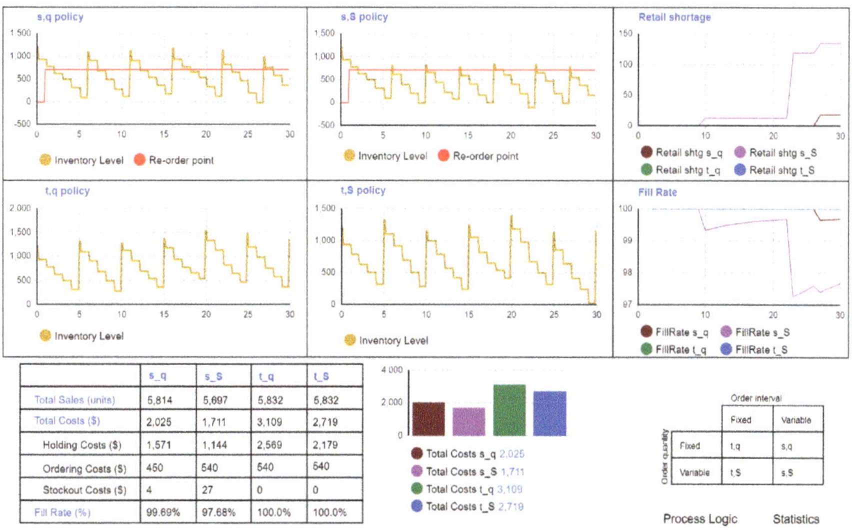


| | s_q | s_S | t_q | t_S |
|---|---|---|---|---|
| Total Sales (units) | 5,814 | 5,697 | 5,832 | 5,832 |
| Total Costs ($) | 2,025 | 1,711 | 3,109 | 2,719 |
| Holding Costs ($) | 1,571 | 1,144 | 2,569 | 2,179 |
| Ordering Costs ($) | 450 | 540 | 540 | 540 |
| Stockout Costs ($) | 4 | 27 | 0 | 0 |
| Fill Rate (%) | 99.69% | 97.68% | 100.0% | 100.0% |

**Fig. 3.54** Experiment #1: analysis view

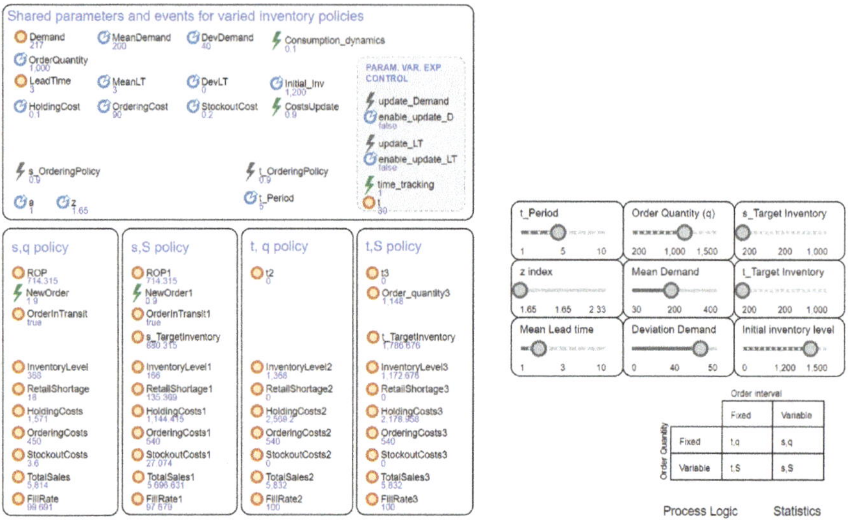


■ **Fig. 3.55** Experiment #1: logic view

Now, assume unpredictable change in demand: after 10 days, mean demand increases from 200 to 250 units since our competitor in the neighborhood is closed for the rest of the month due to unforeseen technical problems.

The simulation results are presented in ■ Fig. 3.56.

It can be observed from ■ Fig. 3.56 that different policies performed differently in regard to the considered demand change. These observations are based on one simulation instance. It would be better to run many replications using a random seed (see Ivanov et al., 2023). ■ Table 3.4 compares the results after 1000 replications.

It can be observed from ■ Fig. 3.56 and Table 4.4 that continuous review policies with dynamic changes in ROP and target inventory allow reacting faster to demand fluctuations.

- **Experiment 2: Impact of Lead Time Dynamics**

The following input data is used (■ Table 3.5).

Simulation results are presented in ■ Fig. 3.57.

It can be observed from ■ Fig. 3.57 that *s*, *S* is the most efficient policy with a cost of \$1503, while periodic review policy *t*, *q* provides 100% service level but at the same time has the highest cost of \$2229.

Now, assume unpredictable change in lead time: after 10 days, mean lead time increases from 3 to 6 days because of a natural disaster that interrupted the transportation route for the rest of the month. This affects continuous review policies only since the supplier in periodic review policies has a period of 7 days between two deliveries.

The simulation results are presented in ■ Fig. 3.58.

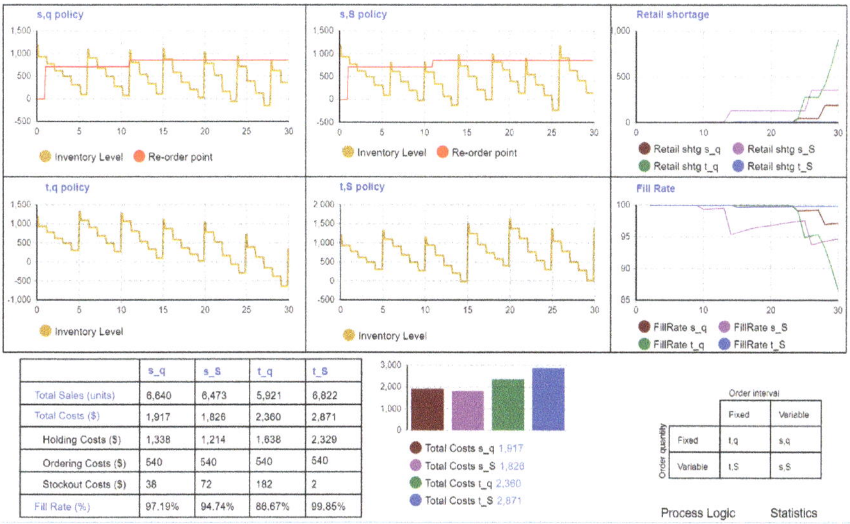


**Fig. 3.56** Experiment #1 with changed demand

**Table 3.4** Result comparison for experiment 1 with and without demand changes

| Policy | Costs | | Fill rate | |
|---|---|---|---|---|
| | No changes in demand | Change in demand | No changes in demand | Change in demand |
| *t*, *q* | 2663 | 2209 | 99 | 75.5 |
| *t*, *S* | 2609 | 2779 | **99.8** | **98.7** |
| *s*, *q* | 1864 | **1798** | 97.1 | 91.1 |
| *s*, *S* | **1734** | 1851 | 93.3 | 93.6 |

It can be observed from Fig. 3.58 that different policies performed differently in regard to the considered lead time change. Table 3.6 compares the results of 1000 replications.

It can be observed from Fig. 3.58 and Table 3.6 that target inventory policies are more robust in regard to lead time fluctuations as compared to order quantity-driven policies.

#### Experiment 3: Impact of Order Quantity

The following input data is used (Table 3.7).

Simulation results are presented in Fig. 3.60.

It can be observed from Fig. 3.59 that *s*, *S* is the most efficient policy with a cost of $1157, but with the worst fill rate at 96.9%.

**Table 3.5** Input data for experiment 2

| Parameter | Description | Value |
|---|---|---|
| OrderQuantity | Order quantity, units | 900 |
| MeanDemand | Mean daily demand, units | 120 |
| DevDemand | Std. dev. of daily demand, units | 40 |
| MeanLT | Lead time, days | 3 |
| DevLT | Std. dev. of lead time | 0 |
| *t* | Time between orders, days | 7 |
| HoldingCost | Holding cost per unit per day | 0.10 |
| OrderingCost | Fixed ordering cost | 90 |
| StockoutCost | Backlogging cost per unit per day | 0.20 |
| t_TargetInventory | Order-up-to inventory target, units | Dynamic |
| s_TargetInventory | Order-up-to inventory target, units | Dynamic |
| *z* | *z*-value for 95% service level | 1.65 |
| Initial | Inv initial inventory level, units | 1000 |
| *a* | Parameter in s_TargetInventory rule | 1 |
| ROP | Reorder point, units | Dynamic |
| ROP1 | Reorder point, units | Dynamic |

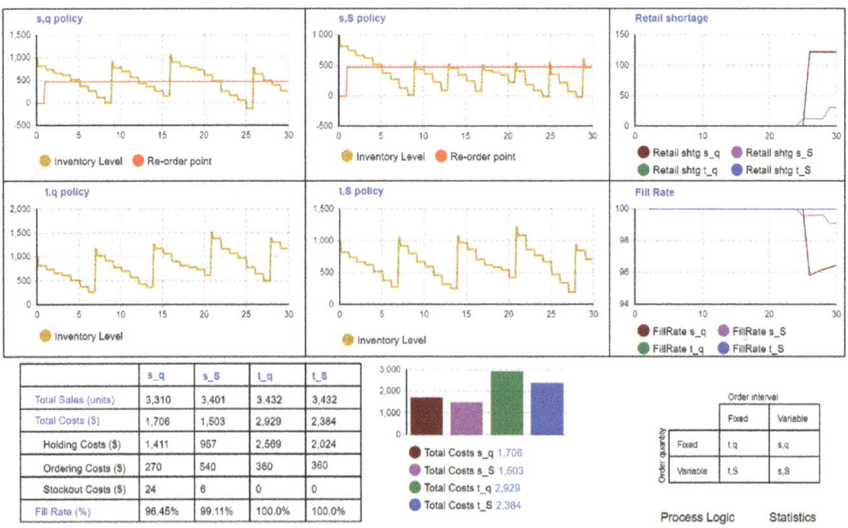


**Fig. 3.57** Experiment #2

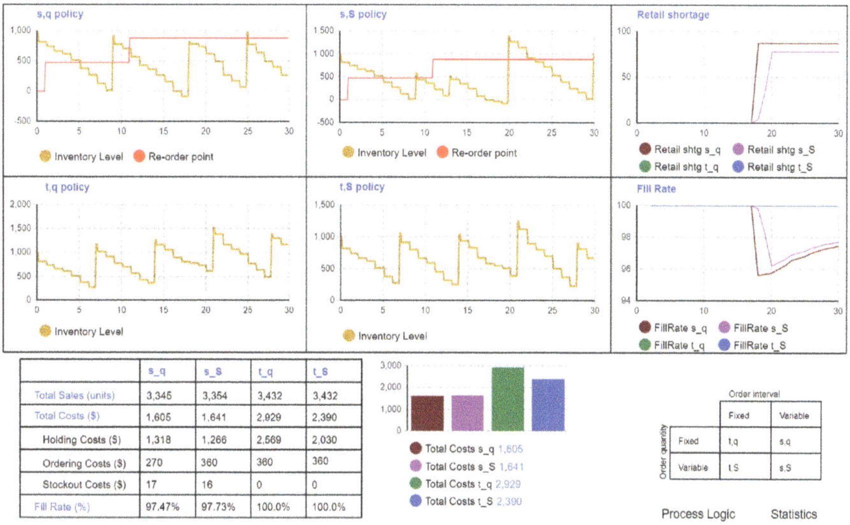


**Fig. 3.58** Experiment #2 with lead time change

**Table 3.6** Result comparison for experiment 2 with and without lead time changes

| Policy | Costs | | Fill rate | |
|---|---|---|---|---|
| | No changes in lead time | Change in lead time | No changes in lead time | Change in lead time |
| *t*, *q* | 2457 | 2454 | 99.6 | **99.6** |
| *t*, *S* | 2266 | 2327 | **99.8** | **99.6** |
| *s*, *q* | 1729 | **1444** | 98.6 | 87.5 |
| *s*, *S* | **1468** | 1759 | 97.5 | 91 |

Let us analyze possible reasons for efficiency and service level of different policies. The following changes can be suggested:

- *t*, *q* policy has too high average inventory on hand → reduce order quantity or increase the reorder period; we will reduce order quantity from 900 to 700 units.
- *t*, *S* policy performs well → no changes.
- *s*, *q* policy has too high average inventory on hand → reduce order quantity or increase the lead time; we will reduce order quantity from 900 to 700 units.
- *s*, *S* policy performs with a shortage → increase the ROP or target inventory; we will increase target inventory as follows: s_TargetInventory = (ROP1-s_S_Inventory) + (**1.5***(MeanDemand + DevDemand)*LeadTime).

The simulation results for the parameters adjusted are presented in Fig. 3.60.

**Table 3.7** Input data for experiment 3

| Parameter | Description | Value |
|---|---|---|
| OrderQuantity | Order quantity, units | 900 |
| MeanDemand | Mean daily demand, units | 80 |
| DevDemand | Std. dev. of daily demand, units | 10 |
| MeanLT | Lead time, days | 3 |
| DevLT | Std. dev. of lead time | 0 |
| *t* | Time between orders, days | 9 |
| HoldingCost | Holding cost per unit per day | 0.10 |
| OrderingCost | Fixed ordering cost | 90 |
| StockoutCost | Backlogging cost per unit per day | 0.20 |
| t_TargetInventory | Order-up-to inventory target, units | 800 |
| s_TargetInventory | Order-up-to inventory target, units | Dynamic |
| *z* | *z*-value for 95% service level | 1.65 |
| Initial | Inv initial inventory level, units | 800 |
| *a* | Parameter in s_TargetInventory rule | 1 |
| ROP | Reorder point, units | Dynamic |
| ROP1 | Reorder point, units | Dynamic |

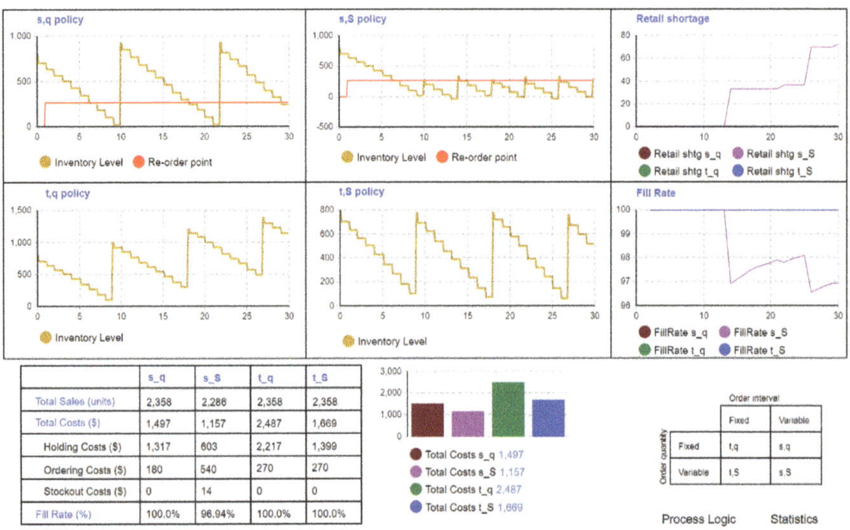


**Fig. 3.59** Experiment #3

It can be observed from ◘ Fig. 3.60 that different policies performed differently in the initial situation and after parameter changes. ◘ Table 3.8 compares the results of 1000 replications.

It can be observed from ◘ Fig. 3.60 and ◘ Table 3.8 that an increase in target inventory positively influenced the *s*, *S* policy in regard to service level while efficiency slightly decreased ($1127 instead of $1123). Here, a trade-off "service level vs. efficiency" can be nicely observed. Nevertheless, *s*, *S* policy still remains the most efficient one. Changes in order quantity also positively influenced the efficiency of *t*, *q* and *s*, *q* policies.

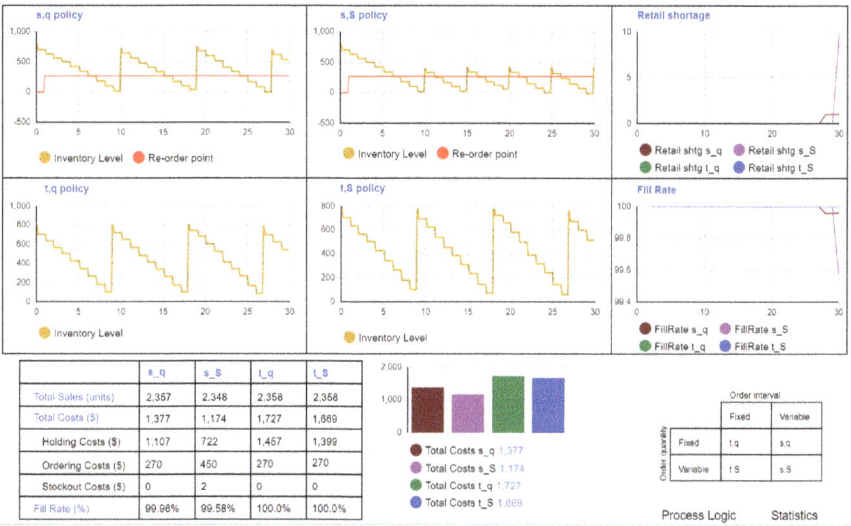


◘ **Fig. 3.60** Experiment #3 with parameter adjustments

◘ **Table 3.8** Result comparison for experiment 3 with and without parameter adjustment

| Policy | Costs | | Fill rate | |
|---|---|---|---|---|
| | No changes | Changes | No changes | Changes |
| *t*, *q* | 2372 | 1613 | **100** | 99.7 |
| *t*, *S* | 1636 | 1637 | 99.9 | **99.9** |
| *s*, *q* | 1432 | 1299 | 98.5 | 97.7 |
| *s*, *S* | **1123** | **1127** | 95.4 | 96.9 |

3

## References

Bozarth, C. C., & Handfield, R. B. (2013). *Introduction to operations and supply chain management* (3rd ed.). Pearson.

Goldratt, E. M. (1990). *What is this thing called theory of constraints and how should it be implemented?* North River Press, Pearson.

Ivanov, D., Tsipoulanidis, A., & Schönberger, J. (2021). *Global supply chain and operations management: A decision-oriented introduction into the creation of value* (3rd ed.). Springer Nature.

Ivanov, D., Millhiser, W., & Nguyen P. (2023). Supply chain and operations simulation with AnyLogic. Retrieved from www.anylogic.com/books

Little, J. D. C., & Graves, S. C. (2008). Little's law. In D. Chhajed & T. J. Lowe (Eds.), *Building intuition: Insights from basic operations management models and principles* (pp. 81–100). Springer.

## Further Reading

Borshchev, A., & Grigoryev, I. (2023). The big book of simulation modeling. Retrieved from https://www.anylogic.com/resources/books/big-book-of-simulation-modeling/

Cachon, G., & Terwiesch, C. (2019). *Matching supply with demand—An introduction to operations management* (4th ed.). McGraw-Hill.

Dolgui, A., & Proth, J.-M. (2010). *Supply chain engineering: Useful methods and techniques*. Springer.

Hopp, W. J., & Spearman, M. L. (2011). *Factory physics* (3rd ed.). Waveland Press.

# Facility Location Planning and Network Optimization

## Contents




D. Ivanov, *Introduction to Supply Chain Analytics*, Classroom Companion: Business,
https://doi.org/10.1007/978-3-031-51241-4_4

**Learning Objectives**

In this chapter, we introduce principles and applications of supply chain analytics to facility location planning and network design. First, we learn how to define the optimal location for a warehouse using the center-of-gravity method. Subsequently, network optimization is introduced. We then explain the software anyLogistix that is comprised of supply chain optimization and simulation analytics. We conclude the chapter with an extended case study integrating facility location planning and network optimization using anyLogistix software.



To this end, our learning objectives for this chapter are as follows:

- Principles and methods of facility location planning and network optimization.
- Center-of-gravity method.
- Network optimization.
- anyLogistix supply chain simulation and optimization software.

## 4.1 Facility Location Planning

### 4.1.1 Problem Statement

A retail company is looking for a new warehouse, from which five major customers should be delivered. Geographical coordinates of the customers and their annual demands are known (◘ Table 4.1).

Our decision problem is defined by customer locations, distances from the warehouse to customers, and customer demands. Each customer location is represented by the ordered pair of $(x; y)$-coordinates. The $(x; y)$-coordinates of the five locations are given in the third as well as the fourth column of ◘ Table 4.1, e.g., $(x_1; y_1) = (10; -80)$. Demand for each customer is also given and denoted as $D(x_i;$

◘ **Table 4.1** Input data

| $i$ | Name | Coordinates | | Annual demand |
|---|---|---|---|---|
| | | $x_i$ | $y_i$ | $D(x_i; y_i)$ |
| 1 | Sport 1-2-3 KG | 10 | −80 | 3 t |
| 2 | TT direct | −45 | −30 | 1 t |
| 3 | Sports and fun | 60 | 50 | 1.5 t |
| 4 | Leisure outlet | 45 | −75 | 3.5 t |
| 5 | Racquets and more | −75 | 80 | 2 t |

$y_i$). Both coordinates and the demand are called *parameters*, i.e., input data for problem solution. In contrast, the ($x$; $y$)-coordinates ($p_x$; $p_y$) of the new warehouse are unknown and have to be determined. Consequently, $p_x$ as well as $p_y$ are the *decision variables* in the investigated decision scenario.

The task is to identify the optimal location of a warehouse so that the total transportation costs from the warehouse to all the customers would be as low as possible. We assume that the total transportation cost sum is proportional to the distance and the transportation volume (i.e., the demand). This leads us to the formulation of the *objective function*, as shown in Eq. (4.1):

$$Z\left(p_x;p_y\right)=\sum_{i-1}^{N}d\left(\left(p_x,p_y\right);\left(x_i,y_i\right)\right)\cdot D\left(x_i,y_i\right)\rightarrow \min \tag{4.1}$$

To minimize the total transportation costs, it is necessary to vary $p_x$ as well as $p_y$ as long as $Z(p_x; p_y)$ becomes minimal. The minimization of (4.1) represents the decision situation of identifying the optimal warehouse location. We call the pair ($p_x$; $p_y$) that minimizes (4.1) an *optimal solution*.

We can observe that the total transportation costs depend on the coordinates $p_x$ and $p_y$ of the prospective warehouses and distances. Since we assumed that the total transportation cost sum from the prospective warehouse location ($p_x$; $p_y$) to a customer location ($x_i$; $y_i$) is more or less equivalent to the distance and demand, the distance $d((p_x; p_y); (x_i; y_i))$ between the $i$th customer location and the warehouse should be determined to calculate transportation costs.

There are two options for calculating the distances. The first option is to use direct line *distance* (Euclidean distance), and the second one is the orthographic (square) distance calculation. According to the first option, distance equals the Euclidean distance between these two points in the plane. This value can be determined according to Pythagoras' theorem leading to Eq. (4.2):

$$d\left(\left(p_x;p_y\right);\left(x_i;y_i\right)\right)=\sqrt{\left(x_i-p_x\right)^2+\left(y_i-p_y\right)^2} \tag{4.2}$$

The orthographic (square) distance calculation is based on Eq. (4.3):

$$d\left(\left(p_x;p_y\right);\left(x_i;y_i\right)\right)=\left|x_i-p_x\right|+\left|y_i-p_y\right| \tag{4.3}$$

Figure 4.1 illustrates the two methods of distance calculation.

Of course, in real life, distances can be computed using digital maps and real routes. We will consider this later in this chapter in anyLogistix models.

### 4.1.2 Center-of-Gravity Method

The model for optimal warehouse location is called the *center-of-gravity model* of location analysis (Chopra & Meindl, 2012). Formulas (4.4) and (4.5) are used to calculate optimal coordinates in the case of direct line distance measurement:

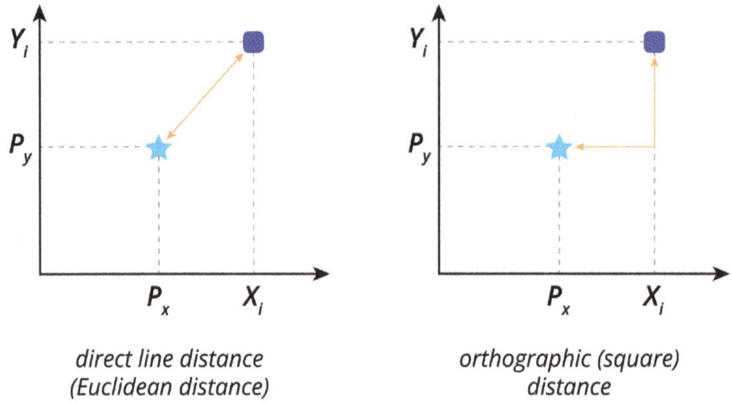


**Fig. 4.1** Distance calculation methods

$$p_x = \frac{\Sigma_{j=1}^{N} \frac{D(x_j;y_j)\cdot x_j}{\sqrt{(p_x - x_j)^2 + (p_y - y_j)^2}}}{\Sigma_{j=1}^{N} \frac{D(x_j;y_j)}{\sqrt{(p_x - x_j)^2 + (p_y - y_j)^2}}} \tag{4.4}$$

$$p_y = \frac{\Sigma_{j=1}^{N} \frac{D(x_j;y_j)\cdot y_j}{\sqrt{(p_x - x_j)^2 + (p_y - y_j)^2}}}{\Sigma_{j=1}^{N} \frac{D(x_j;y_j)}{\sqrt{(p_x - x_j)^2 + (p_y - y_j)^2}}} \tag{4.5}$$

In the orthographic (square) option for distance calculation, we use the rectangular method to determine $p_x$; $p_y$ as shown in Eqs. (4.6) and (4.7):

$$p_x = \frac{\sum_{i=1}^{N} D_i \cdot x_i}{\sum_{i=1}^{N} D_i} \tag{4.6}$$

$$p_y = \frac{\sum_{i=1}^{N} D_i \cdot y_i}{\sum_{i=1}^{N} D_i} \tag{4.7}$$

Using data from Table 4.1, we get $p_x = \frac{30 - 45 + 90 + 157.5 - 150}{11} = 7.5$ and $p_y = \frac{-240 - 30 + 75 - 262.5 + 160}{11} = -27.$

The calculation of distances is summarized in Table 4.2.

Subject to Eq. (4.1), the total cost is calculated as follows assuming that costs per kilometer are $1.00:

**Table 4.2** Distance calculation

| $i$ | Name | Coordinates | | Distances from the point (7.5; −27) | Demand |
|---|---|---|---|---|---|
| | | $x_i$ | $y_i$ | $d(x_i; y_i)$ | $D_i$ |
| 1 | Sport 1-2-3 KG | 10 | −80 | 55.5 | 3 t |
| 2 | TT direct | −45 | −30 | 55.5 | 1 t |
| 3 | Sports and fun | 60 | 50 | 129.5 | 1.5 t |
| 4 | Leisure outlet | 45 | −75 | 85.5 | 3.5 t |
| 5 | Racquets and more | −75 | 80 | 189.5 | 2 t |

$$Z = 55.5 \cdot 3 + 55.5 \cdot 1 + 129.5 \cdot 1.5 + 85.5 \cdot 3.5 + 189.5 \cdot 2 = 1094.5$$

*Note*: In case of different costs per kilometer in different directions, Eqs. (4.4)–(4.7) should be extended. More specifically, the numerators and denominators of the given equations should be multiplied by the transportation costs to each customer individually.

## 4.2 Network Optimization

*Network optimization (NO)* allows to determine optimal locations of several warehouses and factories along with the associated product flows in the supply chain.

We describe a typical NO problem based on Ivanov et al. (2021, Chap. 7.3) (Fig. 4.2).

Let $S$ denote the set of facilities (factories, warehouses, suppliers) that can be installed or used, and let $M$ be the set of all markets to be served. The set $T: = S \times M$ contains all possible transportation links between buyers and suppliers. If a facility $s \in S$ is opened, then the annual costs rise by the amount $f_s$. The corresponding decision is coded again by the binary decision variables $y_s$ ($s \in S$), which equals 1 if a facility is open, and 0 otherwise. The decision to use the transportation link $(s, m) \in T$ between the facility $s \in S$ and the market $m \in M$ increases the annual costs by the additional amount $c_{sm}$. In the context of the NO, it is also necessary to determine the quantity shipped along a transportation link. Therefore, we use the family of nonnegative decision variables $z_{sm}$ to represent decisions about the installation and usage of transportation links. If $z_{sm}$ equals 0, then there is no link installed to connect supplier $s$ with market $m$. If $z_{sm}$ is larger than 0, then the value

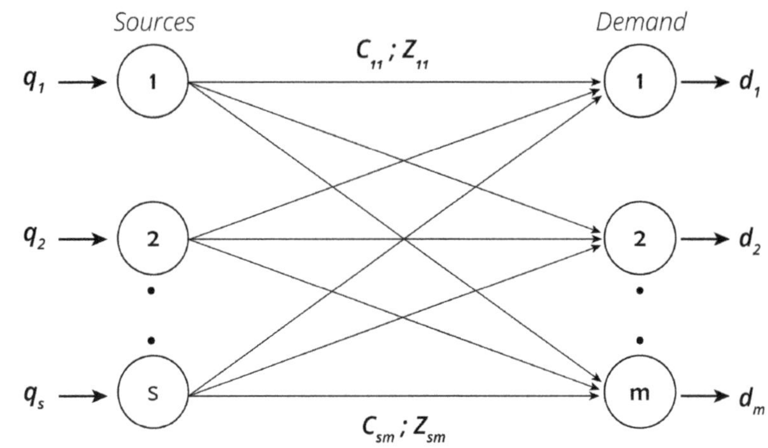


**Fig. 4.2** Network optimization problem

$z_{sm}$ is interpreted as the quantity shipped along the transportation link originating from a supplier $s$ and terminating in a market $m$.

We use the annual fixed cost value $c_{sm}$ to determine the annual costs for the incorporation of the transportation link between a supply location $s$ and market $m$. This parameter determines the annual costs for moving the whole demand of the target market $m$ from the source $s$. In the NO, we aim to distribute the total demand of a market among different transportation links; it is necessary to know the cost $c'_{sm}$ for moving one package unit along the transportation link, connecting supply source $s$ with market $m$.

Using the transportation cost coefficients per package unit as well as the decision variables $z_{sm}$ for the shipped quantities, we can determine the overall distribution cost $Z^{\text{CWLP}}$ per year according to Eq. (4.8):

$$Z^{\text{CWLP}} = \sum_{s \in S} f_s \cdot y_s + \sum_{s \in S} \sum_{m \in M} c'_{sm} \cdot z_{sm} \tag{4.8}$$

Note that the abbreviation CWLP stands for Capacitated Warehouse Location Problem and refers to the class of network optimisation problems that we consider in this section. Using the flow quantity variable $z_{sm}$, we define the demand coverage constraint (4.9), so that a supply of a market from more than one source also becomes possible:

$$\sum_{s \in S} z_{sm} \geq d_m, \quad \forall m \in M \tag{4.9}$$

Similarly, we have to ensure that no supplier intends to deliver more quantities than available to all markets together. The left side of constraint (Eq. 4.10) calculates the actual supply quantities delivered from supplier $s$ to all other markets, and this quantity must not exceed the available quantity $q_s$:

$$\sum_{m \in M} z_{sm} \leq q_s, \quad \forall s \in S \tag{4.10}$$

Again, it is necessary to ensure that a facility $s$ sends out quantities only if this facility $s$ is opened, i.e., it is necessary to code the implication $z_{sm} > 0 \Rightarrow y_s = 1$ in a linear constraint. In the constraint (Eq. 4.11), $K$ represents a "sufficiently large" number. If $z_{sm} > 0$ for a certain market $m$ and if Eq. (4.11) should be valid, then it is necessary that $K \cdot y_s > 0$, which is equivalent to the fact that $y_s > 0$, and this means that supplier $s$ is considered. This technique to model logical implications between nonbinary and *binary decision variables* is called the "big-$K$ method" or "*big-M method*." It is necessary to select an appropriate value for $K$ for each scenario that is modeled:

$$z_{sm} \le K \cdot y_s, \quad \forall s \in S, \quad \forall m \in M \tag{4.11}$$

Finally, the domains of the incorporated decision variables are declared by constraint (4.12):

$$y_s \in \{0;1\} \, \forall s \in S, \quad z_{sm} \ge 0 \forall (s,m) \in T \tag{4.12}$$

Using the aforementioned mathematical expressions, we are now ready to state the NO model as follows:

It is necessary to minimize the total costs for the installation of facilities and the shipment of goods as expressed in Eq. (4.8), so that each market's demand is covered (Eq. 4.9), but no supplier distributes more than its local stock quantity (Eq. 4.10). If we use a facility for supplying a market, then this facility must be open (Eq. 4.11). Each available facility is either opened or closed, and the flows to be shipped between each pair of suppliers and markets must be determined (Eq. 4.12).

A pure formalized problem formulation of the NO is as follows: minimize Eq. (4.8) under the constraints in Eqs. (4.9)–(4.12). The solution of NO problem is a network design and associated production-distribution plan (◘ Fig. 4.3):

Solution of NO problems can be based on exact algorithms (e.g., branch and bound) or heuristics (see Ivanov et al., 2021, Chap. 7.3).

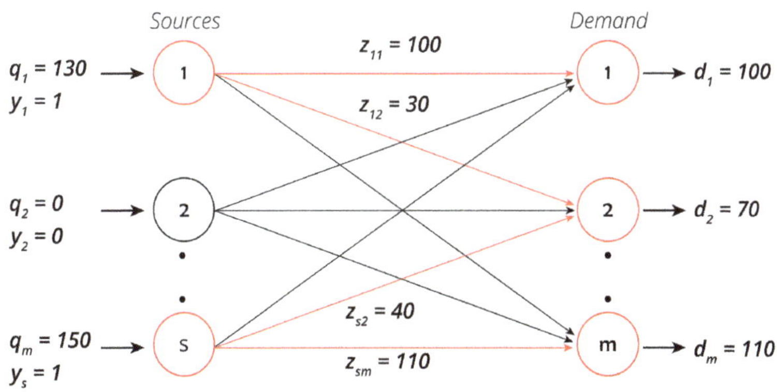


◘ **Fig. 4.3** Example of a solution to network optimization problem

## 4.3 anyLogistix Software

Supply chain network design and operational planning decisions can have a drastic impact on the profitability and success of a company. Whether to have one warehouse or two, close a factory or rent a new one, or to choose one network path over another one are all important decisions a supply chain manager must make. One of the decision support tools is anyLogistix, a software which facilitates greenfield analysis (GFA), NO, and simulation.



*anyLogistix* has become more and more popular with the provision of the free PLE version and because it is an easy-to-use software, includes simulation and optimization, and covers all standard teaching topics (center of gravity, efficient vs. responsive supply chain design, supply chain design through NO, inventory control simulation with safety stock computations, sourcing (single vs. multiple) and shipment (LTL vs. FTL) policy simulation, and milk-run optimization). The following themes will be considered in the anyLogistix case studies in this book:

- Facility location planning (center-of-gravity method)
- Supply chain design (NO)
- Inventory control policies (simulation, safety stocks, ordering policies)

This collection of exercises is based on the main anyLogistix Handbook and Educational Toolkit, which provide technical descriptions of how to build anyLogistix models and are available at ► www.anylogistix.com.

The case studies are designed to stimulate and enhance conceptual and analytical decision-making skills in actual operating situations. Using the models available in anyLogistix, we will conduct analyses to (1) determine an optimal location using GFA for a new warehouse, given the location of their current customers and those customers' relative demands (see ► Sect. 4.1); (2) compare alternative network designs using NO (see ► Sect. 4.2); (3) perform a simulation of different scenarios; (4) validate the models using validation and comparison experiments; and (5) analyze supply chain behavior under uncertainty and disruptions.

We can model the supply chain in two ways (◘ Fig. 4.4):

- Analytical modeling that uses optimization models
- Simulation modeling

You will usually start the first stage of a project (i.e., a scenario in anyLogistix) at the strategic level by using a greenfield analysis (GFA), known as a center-of-gravity analysis (see ► Sect. 4.1), to define the optimal locations of distribution centers (DCs). At this stage, a high level of abstraction with a minimum number of details is used. Existing data, such as customer locations, demand per customer, number and location of DCs, and/or service distances, are used as inputs to the analysis. The output of the analysis is an approximate, optimal location for a production or warehousing facility at which the cost of all in- and outbound transportation is minimized.

During the second stage—the NO—you will extend the problem setting by including feasible facility locations and use other parameters, such as fixed facility

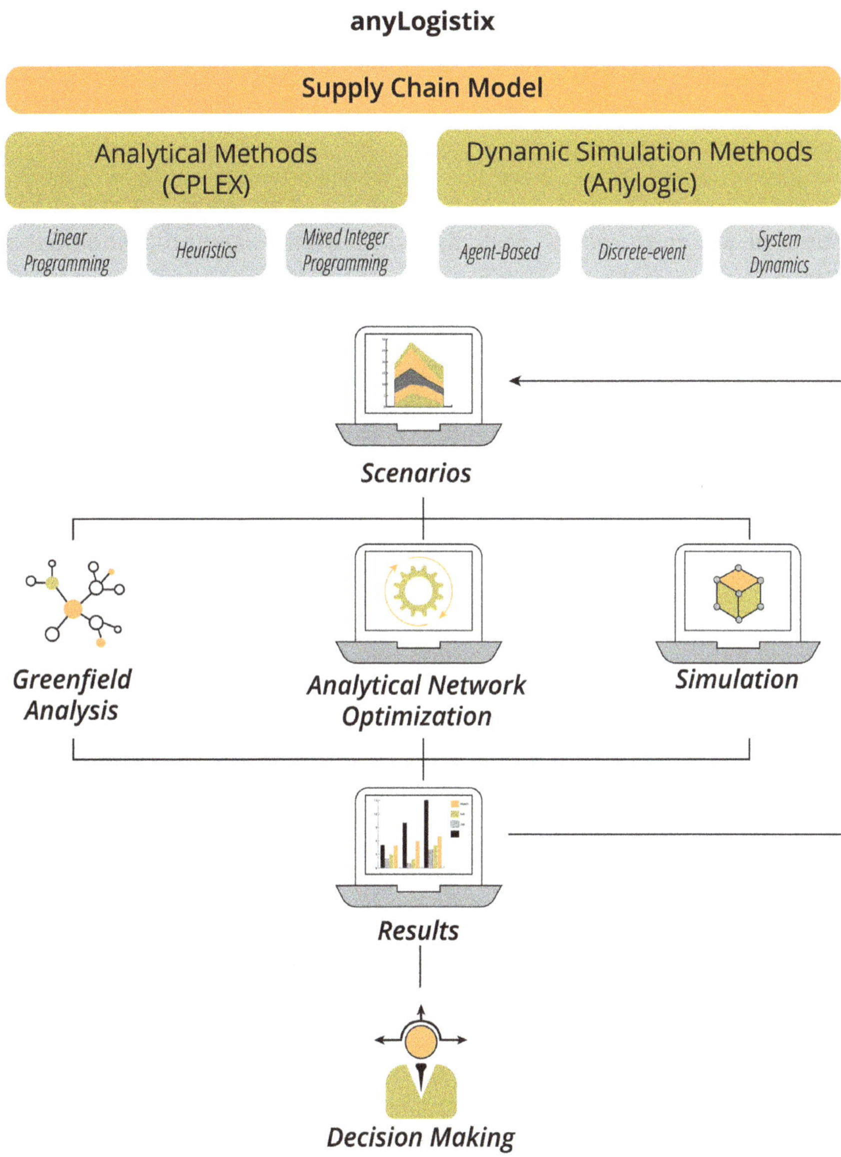


■ **Fig. 4.4** Analytical and simulation methods in anyLogistix

costs, inventory carrying costs, facility opening/closure costs, $CO_2$ emissions, and many periods, and perform NO. NO is a decision support quantitative model for SCM, which allows a supply chain manager to easily compare alternative network designs according to a customizable cost objective function. In contrast to the GFA, through an optimization analysis, many alternative network designs and paths can be compared according to their impact on supply chain efficiency. The results also allow the maximal profitability of each potential alternative network design to be compared with one another. However, a real supply chain is complex and subject to uncertainty, and it is difficult to include many time-dependent, dynamic factors in optimization.

As your problem becomes more detailed, we will extend the analysis in the third stage using simulations, which provide an overview of the effects of different combinations of inventory control, sourcing, transportation, and production policies (◘ Fig. 4.5).



Finally, you will use the results of GFA, NO, and simulation for decision-making. In doing so, it will be an important task to validate the results using sensitivity analysis and compare different scenarios subject to some KPIs. This will be done using variation and comparison analysis in anyLogistix. Conducting a *sensitivity analysis* with different iterations, a so-called variation analysis, highlights model reaction to changes in input parameters and provides a check for robustness. This can best be done by altering various key input parameters such as demand, inventory, or costs. The results then show whether any changes will have severe impacts on the network with regard to cost increases and savings decreases.

You can also use AnyLogic's agent-based, discrete event, and system dynamics simulation models to customize inventory control, sourcing, transportation, and production policies as well as distribution centers, customers, and suppliers (see ► Chap. 3).

We think that you will find working with anyLogistix to be intuitive, and you will find helpful descriptions of the program's features throughout this book. Enjoy your supply chain simulation and optimization with anyLogistix!

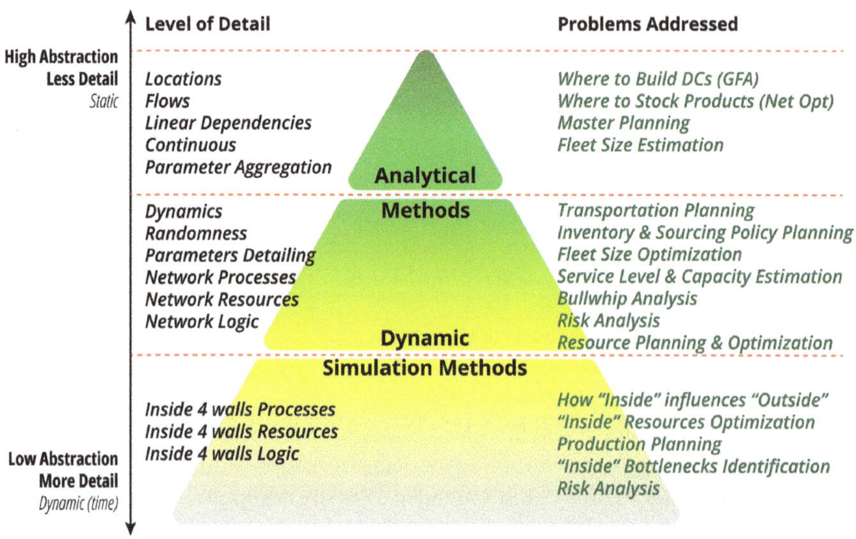


◘ **Fig. 4.5** A pyramid of supply chain design and analysis problems

## 4.4 anyLogistix Case Study

### 4.4.1 Case Description

We consider a company called Polarbear Bicycle. Polarbear Bicycle has been newly founded as an e-commerce start-up selling bicycles. Polarbear's portfolio includes four different types of bicycles: x-cross, urban, all terrain, and tour bicycles. Polarbear needs to find a good location for a new distribution center (DC). First, they estimate customer demand. Polarbear distributes their bicycles to four locations throughout Germany: Cologne, Bremen, Frankfurt am Main, and Stuttgart. ◘ Table 4.3 shows customer demand, which is equal to 245 bicycles per day.

Polarbear Bicycle has hired a consulting firm to analyze supply and distribution network alternatives. They are charged with conducting a GFA to determine the possible location of a new DC in Germany, as well as an NO to compare several options for network designs. The consulting firm was also asked to run a simulation to validate several KPIs and plan inventory and to conduct a sensitivity analysis to verify all the results as well.

◘ **Table 4.3** Customer demand

| Customer | Bicycle type | Demand per day |
|---|---|---|
| Cologne | X-cross | 2 |
| Cologne | Urban | 50 |
| Cologne | All terrain | 15 |
| Cologne | Tour | 10 |
| Bremen | X-cross | 7 |
| Bremen | Urban | 30 |
| Bremen | All terrain | 20 |
| Bremen | Tour | 20 |
| Frankfurt am Main | X-cross | 6 |
| Frankfurt am Main | Urban | 5 |
| Frankfurt am Main | All terrain | 4 |
| Frankfurt am Main | Tour | 5 |
| Stuttgart | X-cross | 15 |
| Stuttgart | Urban | 15 |
| Stuttgart | All terrain | 1 |
| Stuttgart | Tour | 40 |

### 4.4.2 Greenfield Analysis

The objective of the GFA is to determine the best location for our DC. We want to find the location that allows us to fulfill our customer demands at the lowest total transportation cost (see ▶ Sect. 4.1). For GFA, we consider our customers' locations, the distances from our warehouse(s) to our customers, and our customers' demands for our products.



Please install the free version of anyLogistix, PLE version, or anyLogistix Studio version in order to follow the examples considered in this chapter. For technical implementation of the anyLogistix exercises, please consult the anyLogistix Handbook and Educational Toolkit, which provide technical descriptions on how to build anyLogistix models and are available at ▶ www.anylogistix.com. All the models we used in this book are available in the Educational Toolkit, which you can download at ▶ https://www.anylogistix.com/academic/. In addition, all these models are available at the companion website of this book.

Now, we conduct a GFA for the outbound network of Polarbear Bicycle considering the four customers located in Cologne, Bremen, Frankfurt am Main, and Stuttgart. The aim of this GFA is to determine the optimal location of one new DC in Germany subject to total minimum transportation costs.

#### ■ Creating an anylogistix Model

Step 1: Open scenario PB GFA Level 1.

Step 2: Check the tables Customers, Demand, and Products. The data in these tables should correspond to ◘ Table 4.3.

#### ■ Performing Experiments

Step 1: Go to GFA Experiment and run it for "Number of sites = 1" and the period of 1 year.

Step 2: Analyze the results using statistics "Flows" and "New Sites":

(a) What are the optimal coordinates of the DC?
(b) What is the maximum distance from the optimal DC location to a customer?
(c) What is the minimum distance from the optimal DC location to a customer?
(d) What are the total costs of the supply chain? (Note: To compute the sum of costs or flows in GFA results, just slightly drag the heading of the column "Period" or "From" in table "Product flows" in the space over the table.)
(e) Compare the data in statistics "Flows" and table "Demand." Do we satisfy all customer demand from the optimal DC location?
(f) What other costs were not considered in selecting the optimal facility location in the GFA?

#### ■ Solution

Step 1: Go to GFA Experiment and run it for "Number of sites = 1" and the period of 1 year (◘ Fig. 4.6).

Step 2: Analyze the results using statistics "Flows" and "New Sites":

(a) What are the optimal coordinates of the DC?
50.82; 7.47

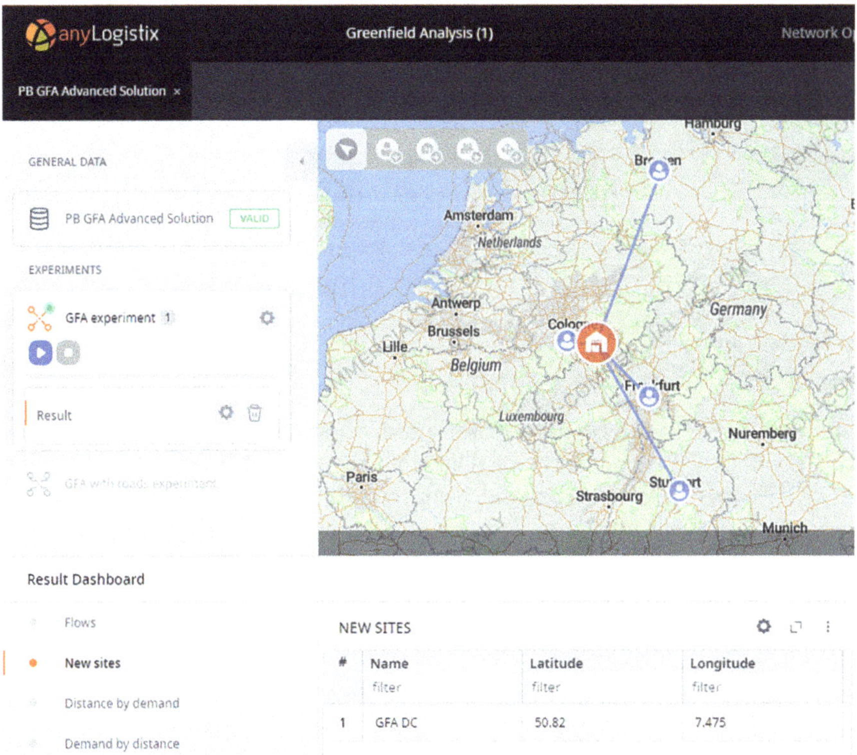


**Fig. 4.6** GFA experiment results

(b) What is the maximum distance from the optimal DC location to a customer?
266.11 km to Bremen

(c) What is the minimum distance from the optimal DC location to a customer?
43.55 km to Cologne

(d) What are the total costs of the supply chain?
$16,351,067

(e) Compare the data in statistics “Flows” and table “Demand.” Do we satisfy all customer demand from the optimal DC location?
Yes, total flows equal total demand. (We compare data in table “Demand” and flows in the statistics “Flows.”)

(g) What other costs were not considered in selecting the optimal facility location in the GFA?
Fixed facility costs, inventory holding costs, processing costs.

### 4.4.3 Network Optimization

After selling the bicycles from the newly established DC according to the GFA results, Polarbear decided to produce their own bicycles. Their production facility has been established in Nuremberg, and 250 bikes are produced each day. Recently,

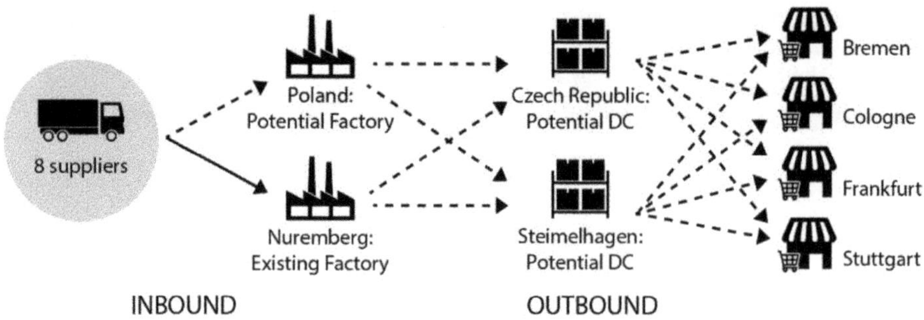


**Fig. 4.7** NO alternatives

they have received an offer from a Polish production factory to rent a DC in the Czech Republic at a reasonable price. The same company also wants to offer them rental of a factory in Warsaw, Poland, even though they already have one factory in Germany. Polarbear must now decide which supply chain design is more profitable:

- Option 1: DC in Germany and factory in Germany
- Option 2: DC in Germany and factory in Poland
- Option 3: DC in the Czech Republic and factory in Poland
- Option 4: DC in the Czech Republic and factory in Germany

In Fig. 4.7, different possibilities for the network design are shown. The dotted lines show possible alternatives and the solid lines, the existing structure of Polarbear's supply chain.

The aim of the NO is to determine which network design is optimal based on Polarbear's selected KPIs, e.g., profit.

Therefore, the factory in Warsaw in Poland, the DC in the Czech Republic, and the DC in Steimelhagen were added as inputs to the model along with the Nuremburg factory. To enable the model's calculation, the reality of the case must be simplified: all demand is assumed to be deterministic without any uncertain fluctuations. To define the two-stage NO problem (transportation between factories and DCs and between DCs and customers) from a mathematical perspective, several parameters must be considered (Table 4.4).

The costs of the rent for the factory in Poland and the DC in the Czech Republic are included in "other costs." For transportation, we assume that each truckload fits 80 bicycles, and trucks travel at a speed of 80 km/h.

- **Creating an anyLogistix Model**

Step 1: Open scenario PB NO Level 1.

Step 2: Check data from Table 4.4 in tables "DCs and Factories," "Facility Expenses," "Paths," "Processing Costs," "Product Flows," "Product Storages," "Production," "Products," and "Vehicle Types." Explain the data.

**Table 4.4** Cost inputs to optimization model

| Costs | Value in USD |
|---|---|
| Factory Nuremberg: fixed (other) costs, per day | 15,000 |
| Factory Poland: fixed (other) costs, per day | 5000 |
| DC Germany: fixed (other) costs, per day | 15,000 |
| DC Germany: carrying costs (per bicycle) | 3.00 |
| DC Czech Republic: fixed (other) costs, per day | 5000 |
| DC Czech Republic: carrying costs | 2.00 |
| DC Germany: processing costs (inbound and outbound shipping per pcs) | 2.00 |
| DC Czech Republic: processing costs (inbound and outbound shipping per pcs) | 1.00 |
| Factory Nuremberg: production costs (per bicycle) | 250 |
| Factory Poland: production (per bicycle) | 150 |
| All bicycles: product purchasing costs | 30 |
| Transportation costs; paths: from factory to DCs | 0.01 × product(pcs) × distance |
| Transportation costs; paths: from DCs to customers | 0.01 × product(pcs) × distance |
| Unit revenue | 499 |

### ■ Performing Experiments

Step 1: Go to NO Experiment and run it with the Demand variation type "95–100%."

Step 2: Analyze the results using statistics "Optimization Results," "Flow Details," "Production Flows," "Demand," and "Overall Stats":

(a) What is the most profitable supply chain design?
(b) Is demand for all customers satisfied?
(c) What is the total revenue of the most profitable supply chain?
(d) What is the total profit of the most profitable supply chain?
(e) Compare the data in statistics "Production Flows" and table "Demand." Does the production quantity correspond to the total demand?
(f) Compare the optimal supply chain design as computed in the NO and the initial supply chain design (factory and DC in Germany) in terms of profit.
(g) What other costs should be considered when redesigning the supply chain according to NO results?
(h) What other factors, apart from costs, should be considered when redesigning the supply chain according to the results of the NO?

## ▪ Solution

Step 1: Go to NO Experiment and run it.

Note: In order to run the NO experiment, you might need to change the units in experiment settings from m$^3$ to pcs to align it with product data

Step 2: Analyze the results using statistics "Optimization Results," "Flow Details," "Production Flows," "Demand," and "Overall Stats":

(a) What is the most profitable supply chain design?

See statistics "Optimization Results": Factory in Poland, DC in the Czech Republic (◘ Fig. 4.8).

(b) Is demand for all customers satisfied?

Yes. See statistics "Demand" and columns "Satisfied" and "Percentage" (◘ Table. 4.5).

(c) What is the total revenue of the most profitable supply chain?

$44,623,075.0; see statistics "Overall Stats."

(d) What is the total profit of the most profitable supply chain?

$26,338,905.352; see statistics "Overall Stats."

(e) Compare the data in statistics "Production Flows" and table "Demand." Does the production quantity correspond to the total demand?

Yes. See statistics "Production Flows."

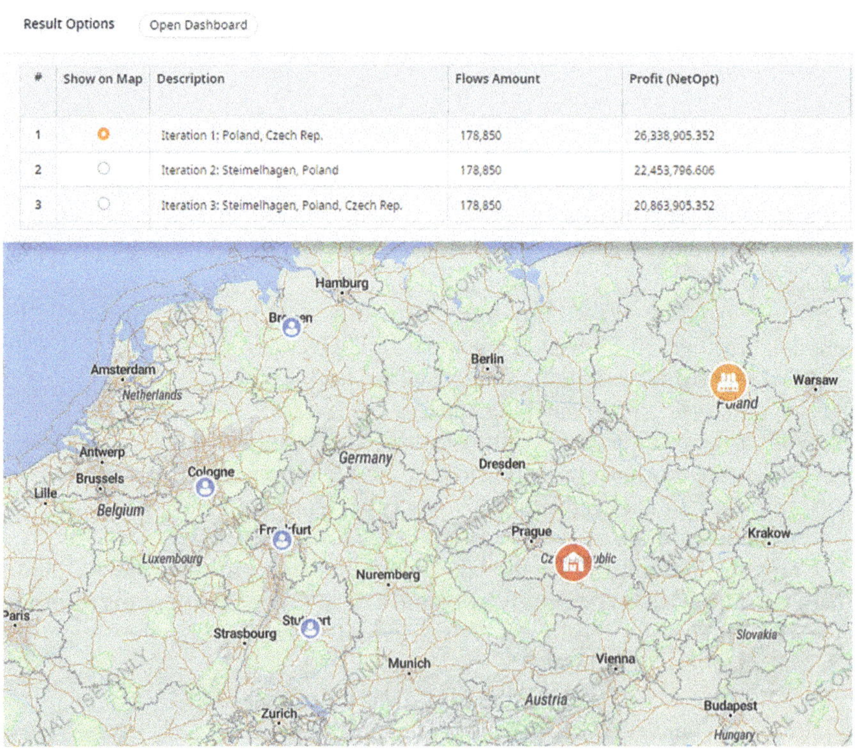


**◘ Fig. 4.8** NO results

(f) Compare the optimal supply chain design as computed in the NO and the initial supply chain design (factory and DC in Germany) in terms of profit.

The initial supply chain profit is shown in Iteration 8 in ◘ Fig. 4.7, and the optimal supply chain profit is shown in Iteration 1. The profit can be increased from $10,394,880.062 to $26,338,905.352 by changing the supply chain design as shown in ◘ Fig. 4.9.

The supply chain redesign would yield an increase in net profit for Polarbear, which is almost three times higher than the approximation of the as-is scenario. This large increase is primarily the result of the cost savings possible by using facilities in Poland and the Czech Republic, which have much lower fixed operating costs.

◘ **Table 4.5** NO results (Iteration 1 and Basic period)

| Customer | Product | Demand min | Demand max | Satis-fied | Percent-age | Revenue, total |
|---|---|---|---|---|---|---|
| Frankfurt am Main | Bicycle "urban" | 1733.75 | 1825 | 1825 | 100 | 910,675 |
| Frankfurt am Main | Bicycle "x-cross" | 2080.5 | 2190 | 2190 | 100 | 1,092,810 |
| Frankfurt am Main | Bicycle "tour" | 1733.75 | 1825 | 1825 | 100 | 910,675 |
| Frankfurt am Main | Bicycle "all terrain" | 1387 | 1460 | 1460 | 100 | 728,540 |
| Stuttgart | Bicycle "urban" | 5201.25 | 5475 | 5475 | 100 | 2,732,025 |
| Stuttgart | Bicycle "x-cross" | 5201.25 | 5475 | 5475 | 100 | 2,732,025 |
| Stuttgart | Bicycle "tour" | 13,870 | 14,600 | 14,600 | 100 | 7,285,400 |
| Stuttgart | Bicycle "all terrain" | 346.75 | 365 | 365 | 100 | 182,135 |
| Cologne | Bicycle "urban" | 17,337.5 | 18,250 | 18,250 | 100 | 9,106,750 |
| Cologne | Bicycle "x-cross" | 693.5 | 730 | 730 | 100 | 364,270 |
| Cologne | Bicycle "tour" | 3467.5 | 3650 | 3650 | 100 | 1,821,350 |
| Cologne | Bicycle "all terrain" | 5201.25 | 5475 | 5475 | 100 | 2,732,025 |
| Bremen | Bicycle "urban" | 10,402.5 | 10,950 | 10,950 | 100 | 5,464,050 |
| Bremen | Bicycle "x-cross" | 2427.25 | 2555 | 2555 | 100 | 1,274,945 |
| Bremen | Bicycle "tour" | 6935 | 7300 | 7300 | 100 | 3,642,700 |
| Bremen | Bicycle "all terrain" | 6935 | 7300 | 7300 | 100 | 3,642,700 |

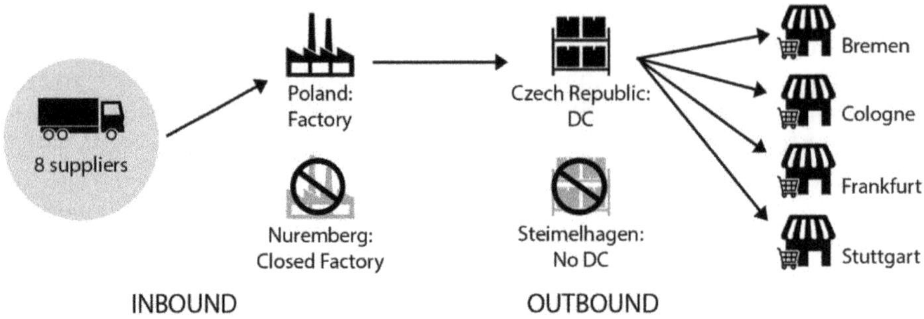


**Fig. 4.9** Optimized network design

(g) What other costs should be considered when redesigning the supply chain according to NO results?
Opening/closure costs
(h) What other factors, apart from costs, should be considered when redesigning the supply chain according to NO results?
Workforce qualification and know-how, disruption risks, future market trends, changes in supplier structures, and risks of outsourcing.

The optimization results show that the highest profit can be achieved in the supply chain design with a DC in the Czech Republic and a factory in Poland. However, the negotiations with the factory in Poland revealed another constraint: the Polish factory would only consider annual quantities of each bicycle type within the range of 10,000 units (minimum capacity utilization) and 25,000 units (maximum capacity utilization). Polarbear must now conduct another NO to include this additional capacity constraint (see Eq. 4.11).

- **Creating an anyLogistix Model**

Step 1: Open scenario PB NO Level 1 Constrained.

Step 2: Check data in table "Production" in the columns "Min Throughput" and "Max Throughput."

- **Performing Experiments**

Step 1: Go to NO Experiment and run it.

Step 2: Analyze the results using statistics "Optimization Results," "Flow Details," "Production Flows," "Demand," and "Overall Stats":
(a) What is the most profitable supply chain design considering the capacity constraint of the factory in Poland?
(b) What is the total profit of the most profitable supply chain?
(c) Compare the optimal supply chain design with the capacity constraint, as computed in the second NO, and the optimal supply chain design without the capacity constraint, as computed in the first NO experiment, in terms of profit.
(d) Which differences between solutions in incapacitated and capacitated scenarios can be observed? Explain.

- **Solution**

(a) What is the most profitable supply chain design considering the capacity constraint of the factory in Poland? See ◘ Fig. 4.10.

The optimal supply chain design now is to have a DC in the Czech Republic and two factories in Germany and Poland.

(b) What is the total profit of the most profitable supply chain?

$19,487,446.413

(c) Compare the optimal supply chain design with the capacity constraint, as computed in the second NO, and the optimal supply chain design without the capacity constraint, as computed in the first NO experiment, in terms of profit.

The profit is reduced from $26,338,905.352 to $19,487,446.413. The optimal supply chain design now is to have a DC in the Czech Republic and two factories in Germany and Poland.

(d) What differences between solutions in incapacitated and capacitated scenarios can be observed? Explain (see ◘ Table. 4.6).

As the "Flow Details" statistics show, the capacity of the factory in Nuremberg will be used to produce the "tour" and "urban" bicycles for which the total demand is higher than the maximum capacity of the factory in Poland. This increases costs (◘ Table. 4.7).

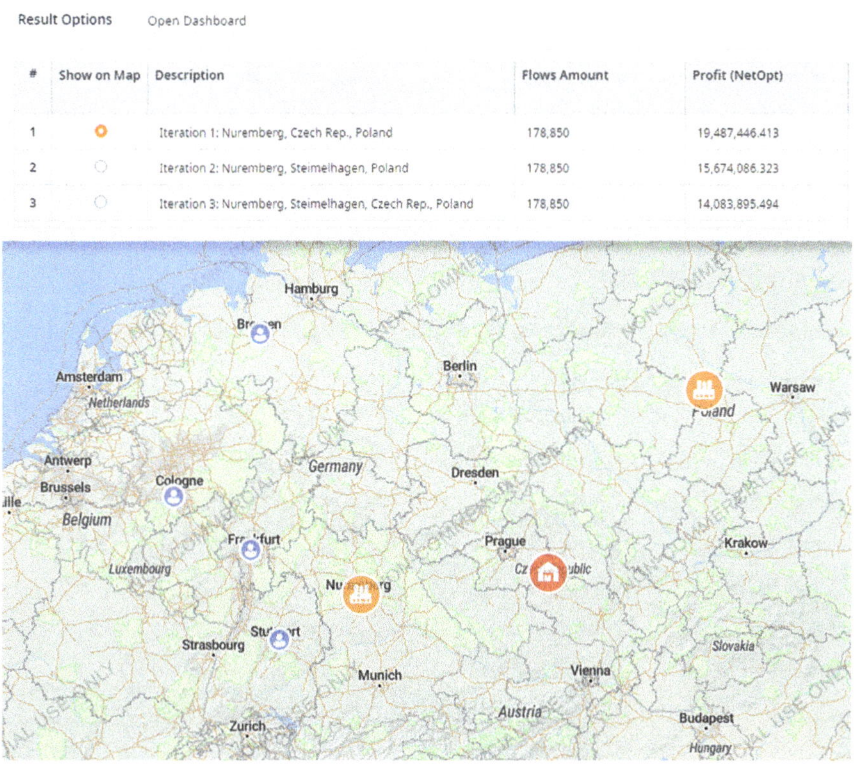


◘ **Fig. 4.10** Optimal network design under constrained capacity

Polarbear's management is interested in the optimization results and starts thinking of redesigning the supply chain. At the same time, the marketing manager of Polarbear expresses concerns that the NO model considers only one period and demand is deterministic. In reality, demand for many products is stochastic and subject to seasonal fluctuations in four periods, i.e., winter, spring, summer, and fall (◘ Tables 4.8 and 4.9).

Polarbear conducts an NO again to include these additional features.



- **Creating an anyLogistix Model**

Step 1: Modify your scenario PB NO Advanced or PB NO Advanced Constrained in tables "Demand," "Periods," and "Production." Assume that the existing capacity constraints for the factory in Poland are distributed equally among the periods.

◘ **Table 4.6** Optimal product flows (Iteration 1 and Basic period)

| From | To | Product | Flow (pcs) |
|---|---|---|---|
| Nuremberg | Czech Rep. | Bicycle "tour" | 2375 |
| Nuremberg | Czech Rep. | Bicycle "urban" | 11,500 |
| Poland | Czech Rep. | Bicycle "all terrain" | 14,600 |
| Poland | Czech Rep. | Bicycle "tour" | 25,000 |
| Poland | Czech Rep. | Bicycle "urban" | 25,000 |
| Poland | Czech Rep. | Bicycle "x-cross" | 10,950 |
| Czech Rep. | Stuttgart | Bicycle "all terrain" | 365 |
| Czech Rep. | Stuttgart | Bicycle "tour" | 14,600 |
| Czech Rep. | Stuttgart | Bicycle "urban" | 5475 |
| Czech Rep. | Stuttgart | Bicycle "x-cross" | 5475 |
| Czech Rep. | Frankfurt am Main | Bicycle "all terrain" | 1460 |
| Czech Rep. | Frankfurt am Main | Bicycle "tour" | 1825 |
| Czech Rep. | Frankfurt am Main | Bicycle "urban" | 1825 |
| Czech Rep. | Frankfurt am Main | Bicycle "x-cross" | 2190 |
| Czech Rep. | Cologne | Bicycle "all terrain" | 5475 |
| Czech Rep. | Cologne | Bicycle "tour" | 3650 |
| Czech Rep. | Cologne | Bicycle "urban" | 18,250 |
| Czech Rep. | Cologne | Bicycle "x-cross" | 730 |
| Czech Rep. | Bremen | Bicycle "all terrain" | 7300 |
| Czech Rep. | Bremen | Bicycle "tour" | 7300 |
| Czech Rep. | Bremen | Bicycle "urban" | 10,950 |
| Czech Rep. | Bremen | Bicycle "x-cross" | 2555 |

**Table 4.7** Total supply chain costs

| Iteration | Objective | Revenue | Total inbound costs | Total transportation costs | Total outbound costs | Total production costs | Total fixed costs | Total penalty |
|---|---|---|---|---|---|---|---|---|
| 1 | 19,487,446.41 | 44,623,075 | −89,425 | −1,030,528.587 | −89,425 | −14,801,250 | −9,125,000 | 0 |
| 2 | 15,674,086.32 | 44,623,075 | −178,850 | −1,015,038.677 | −178,850 | −14,801,250 | −12,775,000 | 0 |
| 3 | 14,083,895.49 | 44,623,075 | −103,300 | −931,329.5063 | −103,300 | −14,801,250 | −14,600,000 | 0 |
| 4 | −31,709,345.47 | 37,699,450 | −75,550 | −868,945.4692 | −75,550 | −11,332,500 | −3,650,000 | −53,406,250 |
| 5 | −35,558,530.05 | 37,699,450 | −151,100 | −917,030.0454 | −151,100 | −11,332,500 | −7,300,000 | −53,406,250 |
| 6 | −37,184,345.47 | 37,699,450 | −75,550 | −868,945.4692 | −75,550 | −11,332,500 | −9,125,000 | −53,406,250 |
| 7 | −432,593,750 | 0 | 0 | 0 | 0 | −6,000,000 | −1,825,000 | −424,768,750 |
| 8 | −438,068,750 | 0 | 0 | 0 | 0 | −6,000,000 | −7,300,000 | −424,768,750 |
| 9 | −1,986,182,435 | 44,623,075 | −89,425 | −970,409.5146 | −89,425 | −22,356,250 | −7,300,000 | −2,000,000,000 |
| 10 | −1,989,605,120 | 44,623,075 | −178,850 | −564,244.9377 | −178,850 | −22,356,250 | −10,950,000 | −2,000,000,000 |

**Table 4.8** Stochastic customer demand

| Customer | Bicycle type | Mean demand per day | Standard deviation of demand per day |
|---|---|---|---|
| Cologne | X-cross | 2 | – |
| Cologne | Urban | 50 | 8 |
| Cologne | All terrain | 15 | 2 |
| Cologne | Tour | 10 | 3 |
| Bremen | X-cross | 7 | – |
| Bremen | Urban | 30 | 6 |
| Bremen | All terrain | 20 | 5 |
| Bremen | Tour | 20 | 3 |
| Frankfurt am Main | X-cross | 6 | – |
| Frankfurt am Main | Urban | 5 | – |
| Frankfurt am Main | All terrain | 4 | – |
| Frankfurt am Main | Tour | 5 | – |
| Stuttgart | X-cross | 15 | – |
| Stuttgart | Urban | 15 | – |
| Stuttgart | All terrain | 1 | – |
| Stuttgart | Tour | 40 | 6 |

**Table 4.9** Demand coefficients

| Period | Coefficient | Time |
|---|---|---|
| Winter | 0.7 | January 1–March 31 |
| Spring | 1.2 | April 1–June 30 |
| Summer | 1.5 | July 1–September 30 |
| Fall | 1 | October 1–December 31 |

■ **Performing Experiments**

Step 1: Go to NO Experiment and run it.

Step 2: Analyze the results using statistics "Optimization Results," "Flow Details," "Storage by Products," "Demand," and "Overall Stats":

(a) Explain the inventory allocation in table "Storage by Products."
(b) Explain the values of Demand Min and Demand Max in table "Demand."
(c) Compare the scenario with only a single period and deterministic demand and the new scenario according to each scenario's financial performance in table "Overall Stats."

■ **Solution**

*Creating an anyLogistix Model*

Step 1: Modify your scenario PB NO Advanced Constrained in tables "Demand," "Periods," and "Production." Assume that the existing capacity constraints for the factory in Poland are equally distributed among the periods (◘ Table. 4.10, 4.11, 4.12).

*Performing Experiments*

Step 1: Go to NO Experiment and run it (◘ Fig. 4.11).

Step 2: Analyze the results using statistics "Optimization Results," "Flow Details," "Storage by Products," "Demand," and "Overall Stats" (◘ Table 4.13).

(a) Explain the inventory allocation in table "Storage by Products."

For example, in "Summer," the factory in Nuremberg has an ending inventory of 650 "tour" bicycles. This inventory is also displayed in the next period "Fall" as "Initial Storage." The rationale for this result is the optimal allocation of capacities, fixed costs, and transportation costs in different periods. Since we do not consider any inventory holding costs at the factory in Nuremberg, it is not problematic to use inventory at this factory. If we would introduce carrying costs in table "Facility Expenses" for factories in Nuremberg and Poland, the resulting flows and inventories would be different.

(b) Explain the values of Demand Min and Demand Max in table "Demand" (◘ Table 4.14).

Since we experimented with the setting "95–100%" demand, we can see demand fulfillment estimations for minimum (95%) and maximum (100%) demand.

(c) Compare the scenario with only a single period (your first NO experiment) and deterministic demand and the new scenario according to each scenario's financial performance in table "Overall Stats" (◘ Tables 4.7 and 4.15).

We can observe higher revenue and profits in the new scenario. This is because there is an increase in demand due to higher demand coefficients in different periods and the introduction of demand variability.

**Table 4.10** Stochastic customer demand setting in table Demand for all periods

| Customer | Product | Demand type | Parameters | | | | | | |
|---|---|---|---|---|---|---|---|---|---|
| Cologne | Bicycle "all terrain" | PeriodicDemand | First occurrence | First day | Order interval, days | Normal | Sigma Mean | 2 | 15 |
| Cologne | Bicycle "tour" | PeriodicDemand | First occurrence | First day | Order interval, days | Normal | Sigma Mean | 3 | 10 |
| Cologne | Bicycle "urban" | PeriodicDemand | First occurrence | First day | Order interval, days | Normal | Sigma Mean | 8 | 50 |
| Cologne | Bicycle "x-cross" | PeriodicDemand | First occurrence | First day | Order interval, days | Value | Value | 2 | |
| Bremen | Bicycle "all terrain" | PeriodicDemand | First occurrence | First day | Order interval, days | Normal | Sigma Mean | 5 | 20 |
| Bremen | Bicycle "tour" | PeriodicDemand | First occurrence | First day | Order interval, days | Normal | Sigma Mean | 4 | 20 |
| Bremen | Bicycle "urban" | PeriodicDemand | First occurrence | First day | Order interval, days | Normal | Sigma Mean | 6 | 30 |
| Bremen | Bicycle "x-cross" | PeriodicDemand | First occurrence | First day | Order interval, days | Value | Value | 7 | |
| Frankfurt am Main | Bicycle "all terrain" | PeriodicDemand | First occurrence | First day | Order interval, days | Value | Value | 4 | |
| Frankfurt am Main | Bicycle "tour" | PeriodicDemand | First occurrence | First day | Order interval, days | Value | Value | 5 | |
| Frankfurt am Main | Bicycle "urban" | PeriodicDemand | First occurrence | First day | Order interval, days | Value | Value | 5 | |
| Frankfurt am Main | Bicycle "x-cross" | PeriodicDemand | First occurrence | First day | Order interval, days | Value | Value | 6 | |
| Stuttgart | Bicycle "all terrain" | PeriodicDemand | First occurrence | First day | Order interval, days | Value | Value | 1 | |
| Stuttgart | Bicycle "tour" | PeriodicDemand | First occurrence | First day | Order interval, days | Normal | Sigma Mean | 6 | 40 |
| Stuttgart | Bicycle "urban" | PeriodicDemand | First occurrence | First day | Order interval, days | Value | Value | 15 | |
| Stuttgart | Bicycle "x-cross" | PeriodicDemand | First occurrence | First day | Order interval, days | Value | Value | 15 | |

**Table 4.11** Demand coefficients setting in table Periods

| Name | Start | End | Demand coefficient |
|---|---|---|---|
| Winter | 1/1/21 00:00:00 | 3/31/21 00:00:00 | 0.7 |
| Spring | 4/1/21 00:00:00 | 6/30/21 00:00:00 | 1.2 |
| Summer | 7/1/21 00:00:00 | 9/30/21 00:00:00 | 1.5 |
| Fall | 10/1/21 00:00:00 | 12/31/21 00:00:00 | 1 |

**Table 4.12** Capacity constraint setting in table Production

| Site | Product | BOM | Production cost | Currency | Min throughput | Max throughput | Down penalty | Up penalty |
|---|---|---|---|---|---|---|---|---|
| Nuremberg | Bicycle "all terrain" | | 250 | USD | 0 | 0 | 0 | 0 |
| Nuremberg | Bicycle "tour" | | 250 | USD | 0 | 0 | 0 | 0 |
| Nuremberg | Bicycle "urban" | | 250 | USD | 0 | 0 | 0 | 0 |
| Nuremberg | Bicycle "x-cross" | | 250 | USD | 0 | 0 | 0 | 0 |
| Poland | Bicycle "all terrain" | | 150 | USD | 2500 | 6250 | 50000 | 50000 |
| Poland | Bicycle "tour" | | 150 | USD | 2500 | 6250 | 50000 | 50000 |
| Poland | Bicycle "urban" | | 150 | USD | 2500 | 6250 | 50000 | 50000 |
| Poland | Bicycle "x-cross" | | 150 | USD | 2500 | 6250 | 50000 | 50000 |

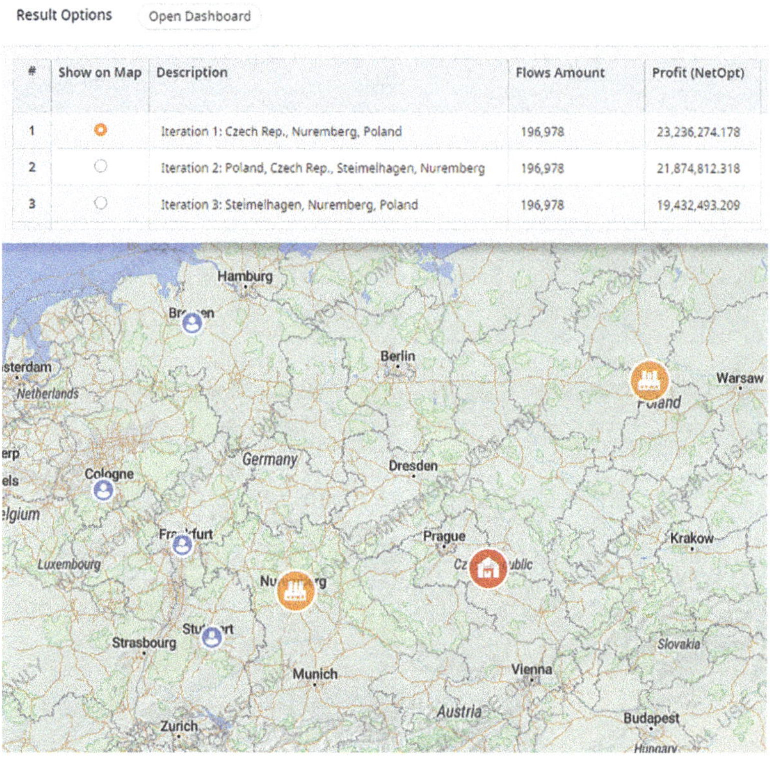


**Fig. 4.11** NO results

**Table 4.13** Storage by Products in Nuremberg (Iteration 1)

| Period | Product | Initial storage | Storage | Out flow |
|---|---|---|---|---|
| Spring | Bicycle "x-cross" | 0 | 0 | 0 |
| Spring | Bicycle "urban" | 0 | 0 | 4670 |
| Spring | Bicycle "all terrain" | 0 | 0 | 0 |
| Spring | Bicycle "tour" | 0 | 0 | 415 |
| Summer | Bicycle "x-cross" | 0 | 0 | 0 |
| Summer | Bicycle "urban" | 0 | 2950 | 7550 |
| Summer | Bicycle "all terrain" | 0 | 0 | 0 |
| Summer | Bicycle "tour" | 0 | 650 | 4100 |
| Fall | Bicycle "x-cross" | 0 | 0 | 0 |
| Fall | Bicycle "urban" | 2950 | 0 | 2950 |
| Fall | Bicycle "all terrain" | 0 | 0 | 0 |
| Fall | Bicycle "tour" | 650 | 0 | 650 |

**Table 4.14** Demand satisfaction analysis (Iteration 1, Period Winter)

| Customer | Product | Demand min | Demand max | Satis-fied | Percent-age |
|---|---|---|---|---|---|
| Frankfurt am Main | Bicycle "x-cross" | 359.1 | 378 | 378 | 100 |
| Frankfurt am Main | Bicycle "urban" | 299.25 | 315 | 315 | 100 |
| Frankfurt am Main | Bicycle "all terrain" | 239.4 | 252 | 252 | 100 |
| Frankfurt am Main | Bicycle "tour" | 299.25 | 315 | 315 | 100 |
| Cologne | Bicycle "x-cross" | 119.7 | 126 | 126 | 100 |
| Cologne | Bicycle "urban" | 2992.5 | 3150 | 3100 | 98.413 |
| Cologne | Bicycle "all terrain" | 897.75 | 945 | 945 | 100 |
| Cologne | Bicycle "tour" | 598.5 | 630 | 630 | 100 |
| Bremen | Bicycle "x-cross" | 418.95 | 441 | 441 | 100 |
| Bremen | Bicycle "urban" | 1795.5 | 1890 | 1890 | 100 |
| Bremen | Bicycle "all terrain" | 1197 | 1260 | 1260 | 100 |
| Bremen | Bicycle "tour" | 1197 | 1260 | 1260 | 100 |
| Stuttgart | Bicycle "x-cross" | 897.75 | 945 | 945 | 100 |
| Stuttgart | Bicycle "urban" | 897.75 | 945 | 945 | 100 |
| Stuttgart | Bicycle "all terrain" | 59.85 | 63 | 63 | 100 |
| Stuttgart | Bicycle "tour" | 2394 | 2520 | 2520 | 100 |

**Table 4.15** Total supply chain costs (PB NO Advanced Extended)

| Itera-tion | Objective | Revenue | Total penalty | Total production costs | Total transportation costs | Total fixed costs | Total inbound costs | Total outbound costs |
|---|---|---|---|---|---|---|---|---|
| 1 | 23,236,274.18 | 49,146,011 | 0 | −16,806,850 | −1,130,908.822 | −7,775,000 | −98,489 | −98,489 |
| 2 | 21,874,812.32 | 49,146,011 | 0 | −16,806,850 | −1,105,170.682 | −9,155,000 | −102,089 | −102,089 |
| 3 | 19,432,493.21 | 49,146,011 | 0 | −16,806,850 | −1,087,711.791 | −11,425,000 | −196,978 | −196,978 |
| 4 | −61,752,558.9 | 38,998,846 | −84,328,750 | −11,723,100 | −893,246.9032 | −3,650,000 | −78,154 | −78,154 |
| 5 | −63,132,558.9 | 38,998,846 | −84328,750 | −11,723,100 | −893,246.9032 | −5,030,000 | −78,154 | −78,154 |
| 6 | −65,608,680.68 | 38,998,846 | −84,328,750 | −11,723,100 | −943,060.6818 | −7,300,000 | −156,308 | −156,308 |
| 7 | −475,885,250 | 0 | −468,060,250 | −6,000,000 | 0 | −1,825,000 | 0 | 0 |
| 8 | −477,265,250 | 0 | −468,060,250 | −6,000,000 | 0 | −3,205,000 | 0 | 0 |
| 9 | −1984030,179 | 49,170,961 | −2,000,000,000 | −24,634,750 | −1,069,311.525 | −7,300,000 | −98,539 | −98,539 |
| 10 | −1,985,352,883 | 49,170,961 | −2,000,000,000 | −24,634,750 | −966,935.7964 | −8,680,000 | −121,079 | −121,079 |

## References

Chopra, S., & Meindl, P. (2012). *Supply chain management. Strategy, planning and operation* (5th ed.). Pearson.

Ivanov, D., Tsipoulanidis, A., & Schönberger, J. (2021). *Global supply chain and operations management: A decision-oriented introduction into the creation of value* (3rd ed.). Springer Nature.

## Further Reading

Daskin, M. S. (1995). *Network and discrete location: Models, algorithms, and applications*. Wiley.

Ivanov, D. (2023) Supply chain simulation and optimization with anyLogistix. Retrieved from anylogistix.com

Klibi, W., Martel, A., & Guitouni, A. (2010). The design of robust value-creating supply chain networks: A critical review. *European Journal of Operational Research, 203*(2), 283–293.

Melo, M. T., Nickel, S., & Saldanha-da-Gama, F. (2009). Facility location and supply chain management—A review. *European Journal of Operational Research, 196*(2), 401–412.

# Supply Chain Risk and Resilience Analytics

## Contents




D. Ivanov, *Introduction to Supply Chain Analytics*, Classroom Companion: Business,
https://doi.org/10.1007/978-3-031-51241-4_5

**Learning Objectives**

In this chapter, we introduce the principles and applications of supply chain analytics to resilience analysis and stress testing the supply networks. First, we learn the major terminology, principles, and methods of supply chain resilience analytics. Subsequently, we consider a case study for supply chain resilience analysis using anyLogistix software. We conclude the chapter with an introduction to digital supply chain twins using the example of anyLogistix and resilience modeling.

To this end, our learning objectives for this chapter are as follows:
- Supply chain risks and resilience.
- Principles and methods of supply chain resilience analytics.
- anyLogistix supply chain simulation and optimization software,
- Digital supply chain twins.



## 5.1 Risk and Resilience in Supply Chains

### 5.1.1 Supply Chain Risks and the Ripple Effect

Supply chains are exposed to risks leading to different types of supply, production, and logistics disruptions. *Risk management* in supply chains has become one of the most important topics in research and practice over the last decade.

Chopra and Sodhi (2004) categorized potential supply chain risks into nine categories:
- Disruptions (e.g., natural disasters, terrorism, war)
- Delays (e.g., inflexibility of supply source)
- Systems (e.g., information infrastructure breakdown)
- Forecast (e.g., inaccurate forecast, bullwhip effect)
- Intellectual property (e.g., vertical integration)
- Procurement (e.g., exchange rate risk)
- Receivables (e.g., number of customers)
- Inventory (e.g., inventory holding cost, demand, and supply uncertainty)
- Capacity (e.g., cost of capacity)

Tang (2006) distinguishes operational risks (e.g., uncertain costs and uncertain demand) and disruption risks (e.g., natural and man-made disasters or general crises). Christopher (2011) divides the supply chain risks into five groups, i.e., supply, demand, process, control, and environmental risks. Another classification of supply chain risks has been proposed by Ivanov and Dolgui (2019) as follows:
- *Network risks*, i.e., risks imposed by specific types of supply chain networks (e.g., scale-free vs. decentralized networks)
- *Process risks*, i.e., specific parameters of supply chain processes such as inventory and capacity may influence supply chain proneness to disruptions
- *Supplier risks*, i.e., risks induced by specific suppliers, e.g., those located in geographical zones with high probability of natural disasters

**Fig. 5.1** Classification of supply chain risks

In Fig. 5.1, we present a classification of *supply chain risks* based on Ivanov (2022b).

Supply chain risks stem from different areas related to material flows such as supply, demand, production, and logistics process. Along with material flow-related risks, risks arise in financial and information flows, e.g., due to cyberattacks or financial crisis. Moreover, supply chains are subject to macro-risks such as climate changes and shortages of critical resources.

From the perspective of impact severity, different types of risks in the supply chain can be classified into demand, supply, process, and structure areas entailing a generalized separation in operational and disruption risks, also referred to as low-impact-high-frequency (LIHF) and high-impact-low-frequency (HILF) risks (see Fig. 5.2).

LIHF risks of demand and supply uncertainty are related to random uncertainty and business-as-usual situation. Such risks are also known as recurrent or *operational risks* (Tang, 2006; Chopra et al., 2007). Disruption risks are characterized by a very strong and immediate impact on the supply chains since some factories, suppliers, warehouses, and transportation links become temporarily unavailable. Adversely, the resulting material shortages and delivery delays propagate downstream the supply chain, causing the *ripple effect* and performance degradation in terms of revenue, service level, and productivity decreases.

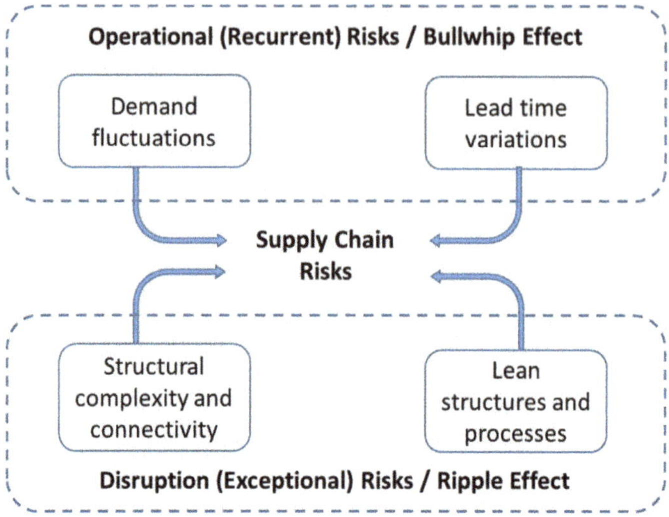


**Fig. 5.2** Operational and disruption supply chain risks (Ivanov, 2021a)

5

The *ripple effect* occurs when a disruption, rather than remaining localized or being contained to one part of the supply chain, cascades downstream and impacts the performance of the entire supply chain (Dolgui et al., 2018). Ripple effect is followed by supply chain structural dynamics (Ivanov, 2018), and its impact might include lower revenues, delivery delays, loss of market share and reputation, etc., with adverse effects on the profitability of supply chain (Li et al., 2021).

According to Dolgui et al. (2020), the ripple effect "refers to structural dynamics and describes a downstream propagation of the downscaling in demand fulfilment in the supply chain as a result of a severe disruption." Ivanov et al. (2014) state: "Ripple effect describes the impact of a disruption on supply chain performance and disruption-based scope of changes in the supply chain structures and parameters." These definitions imply that the ripple effect refers to multi-stage networks and triggering failures in the network elements as a domino or cascading effect.

## 5.1.2 Supply Chain Resilience

*Supply chain resilience* is the "ability to maintain, execute, and recover (adapt) planned execution along with achievement of the planned (or adapted, but yet still acceptable) performance" (Ivanov, 2018). According to Hosseini et al. (2019), supply chain *resilience* is "the firm's capability to withstand, adapt, and recover from disruptions to meet customer demand, ensure target performance, and maintain operations in vulnerable environments."

Consider Fig. 5.3 as an illustration.

Supply chain resilience looks at maintaining some desired performance despite disruptions. Using some proactive capabilities (i.e., inventory), a supply chain can

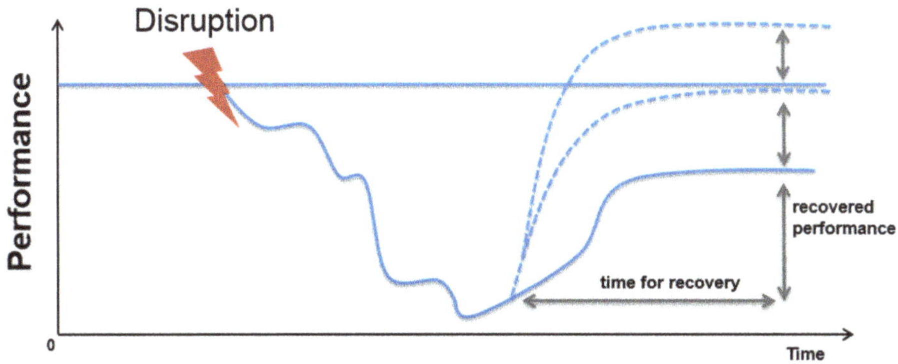


**Fig. 5.3** Supply chain resilience concept (Ivanov, 2022b)

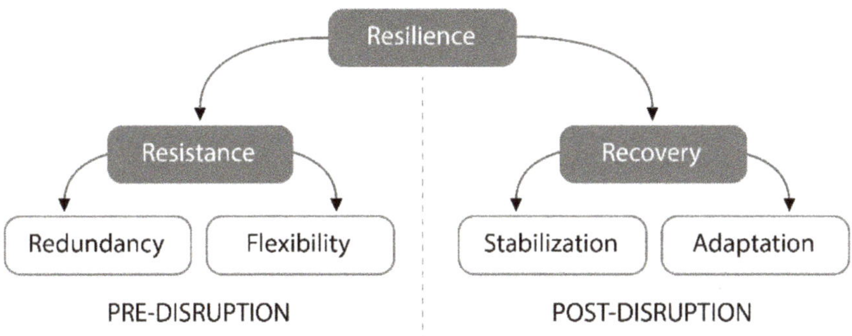


**Fig. 5.4** Supply chain resilience

absorb negative disruption impacts (e.g., supply unavailability) without performance degradation (Hägele et al., 2023). However, if proactive capabilities do not help, performance (e.g., on-time delivery or revenue) can decline. In this case, reactive capabilities should be employed to restore the performance and operations. This takes time and creates costs. Thus far, building a resilient supply chain is based on mitigating risks, preparedness for disruptions, stabilization, and recovery (see Fig. 5.4).

Supply chain resilience capabilities can be classified as two groups, i.e., resistance/redundancy and adaptation/recovery. In *redundancy,* different *reserves* (material inventory, capacities, and network design redundancy) as well as facility fortification can be named. The redundancies are intended to protect the supply chain against disruptions based on certain reserves. This is related to the supply chain resistance or *robustness*.

Supply chain adaptation has four major dimensions, i.e., *scalability* (e.g., online retailers used capacity expansions for coping with surges in demand during the COVID-19 pandemic), *process flexibility* (e.g., auto manufacturers repurposed their production from cars to ventilators), *structural reconfiguration* (e.g., usage of backup suppliers), and *intertwining* (e.g., collaboration of commercial and health-care supply chains) (Ivanov, 2021b). Besides, product substitution can also be used to adapt supply chains. Moreover, organizations can learn from previous disruptions, and so improve future resilience. Adaptation is connected with *flexibility* with regard to both operations and resources.

### 5.1.3 Supply Chain Resilience Analytics

5

Methods for supply chain *resilience analytics* can be classified according to network, process, and control levels (Ivanov & Dolgui, 2021a) as shown in ◘ Fig. 5.5.

The methods at the network level analyze network structures and risk propagations through networks (Li et al., 2020, 2021; Basole & Bellamy, 2014). These methods allow uncovering potential bottlenecks in supply chain designs and detecting disruption scenarios. It becomes possible to identify critical nodes (or combinations of nodes), the failure of which would lead to supply chain discontinuities and operational collapse. Along with the stress testing of existing supply chain designs, the network-level analyses can offer guidance to enhance the resilience, e.g., through supplier diversification. With the use of Bayesian networks, it becomes possible to model dependencies and interdependencies in supply networks with consideration of both vulnerabilities and recovery.

The process-level methods use mathematical optimization to define optimal flow (re)configurations under disruptions and their propagation (Aldrighetti et al., 2023). Optimization models presume some parametrized structures to balance demands, processing capacities, and supply (Sawik, 2020). As the most desirable outcome, process-level analysis seeks to identify and test resilient supply chain designs to sustain disruptions, which range from optimistic and pessimistic scenarios to probability-based disruptions and worst-case scenarios in robust optimization.

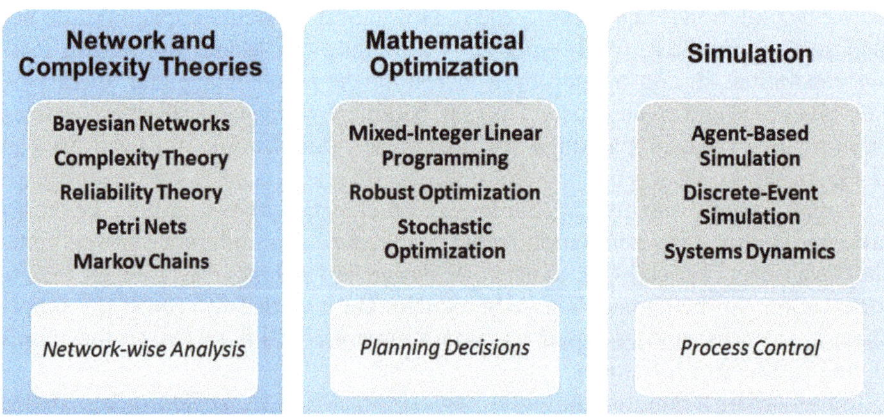


◘ **Fig. 5.5** Modeling methods for supply chain resilience (Ivanov & Dolgui, 2021a)

The control-level studies are distinctively characterized by the inclusion of inventory control and production-ordering dynamics in the analysis. At this level, simulation methods are the most dominant. The control-level methods help to analyze dynamics of disruptions and recovery. Dynamic supply chain behaviors and time dependencies in disruption propagation and responses can be modeled conveniently using simulation techniques. In ◘ Table 5.1, we summarize major methods, their outcomes, and managerial applications.

◘ **Table 5.1** Methods, outcomes, and managerial insights (based on Ivanov & Dolgui, 2021a)

| Level of analysis | Methods | Outcomes | Managerial insights |
|---|---|---|---|
| Network level | Graph theory<br>Complexity theory<br>Entropy<br>Petri nets<br>Bayesian networks<br>Markov chains<br>Reliability theory/ statistical analysis | • Associations between network structures and resilience<br>• Analysis of critical network elements leading to supply chain discontinuities<br>• Modeling of interdependencies in supply chains<br>• State dynamics within supply chain nodes<br>• Assessment of supply chain robustness and resilience to disruptions with considerations of ripple effect | • Identification of disruption scenarios of different severity<br>• Stress testing of supply chain designs<br>• Proneness of specific supply chain designs to disruption risk propagation<br>• Identification of critical suppliers and facilities for maintaining supply chain operations<br>• Selection and proactive enhancements of supply chain designs to sustain certain levels of disruptions<br>• Adaptation of supply chain designs according to environmental changes |
| Process level | Stochastic optimization<br>Robust optimization<br>Linear/ mixed-integer programming | • Optimal reconfigurations of material flows according to disruption scenarios<br>• Impacts of ripple effect and structural dynamics on service level and costs<br>• Optimal reallocation of supply and demand under disruptions | • Stress testing of supply chain production-distribution plans within differently disrupted network designs<br>• Analysis of contingency preparedness plans<br>• Recovery plan selection |

(continued)

**Table 5.1**

| Level of analysis | Methods | Outcomes | Managerial insights |
|---|---|---|---|
| Control level | Optimal control<br>Systems dynamics<br>Agent-based simulation<br>Discrete-event simulation | • Impacts of disruptions on service level, inventory levels, and costs<br>• Time-dependent effect of disruptions on supply chain behaviors and performance in dynamics<br>• Individual behavior of firms in supply chains | • Building resilient supply chains for new, post-pandemic business models<br>• Analysis of disruption propagation in dynamics with consideration of production and inventory control policies<br>• Simulation of operation policies during disruptions, in transition to recovery, and in post-recovery periods |

5

### 5.1.4 Supply Chain Viability

> *Viability* is the ability of a supply chain to maintain itself and survive in a changing environment through a redesign of structures and replanning of performance with long-term impacts (Ivanov & Dolgui, 2020; Ivanov, 2022a; Ivanov et al., 2023a).

Supply chain resilience is usually considered the ability to bounce back once disrupted and recover to a known normal state. In contrast, viability looks at the ability to bounce forward and adapt once disrupted when a normal state cannot be reached at a reasonable time scale, and a search for an unknown "new normal" is required. While resilience is triggered by a disruption, the viability is linked to a supply chain crisis. To this end, resilience is considered a disruption-driven supply chain property related to single, discrete, exceptional disruptive events within a closed system setting. These analyses are performance oriented for some fixed time windows and mostly linear, single-flow-directed supply chain systems.

Viability is concerned with systems and their structural dynamics (Ivanov, 2023d). It considers system evolution (i.e., open system context) through disruption-reaction balancing in the open system context. Speaking in terms of chaos theory, with the supply chain viability, we have a case of searching for new attractors to stabilize the system which was thrown into a crisis. Next, resilience is related to the level of individual supply chains or supply networks, whereas viability is concerned with intertwined supply networks "*that encapsulate entireties of interconnected supply chains, which, in their integrity, secure the provision of society and markets with goods and services*" (Ivanov & Dolgui, 2020). Further, we retain the focus of supply chain resilience towards performance in a fixed time window and in a closed system setting, while supply chain viability considers the structural dynamics of continuous system change in an open system context (Dolgui et al., 2023).

To summarize, two main differences between resilience and viability are as follows (Ivanov et al., 2023b; Ivanov & Keskin, 2023; Ivanov, 2023d):

1. Resilience is the ability to recover *after* a disruption (e.g., an earthquake) and return to an *initial state*. Viability is the ability to operate and keep serving markets/customers with products and services *in the presence* of disruptions and long-term crisis (i.e., the ability to survive in the long term) through adaptation and reconfiguration of *changing states dynamically*.
2. Resilience considers performance of individual supply chains. Viability integrates resilience and sustainability considering a compounded function of economical and societal components and intertwined supply networks.

## 5.2 anyLogistix Case Study

In this section, we illustrate supply chain resilience analysis through descriptive, predictive, and prescriptive analytics using *anyLogistix* software.

### 5.2.1 Performance Dashboard Design

To add KPI to the dashboard, in the dashboard area, select Add Chart, and then use the following screen to select the KPIs and the form (■ Fig. 5.6) the KPIs will take. By default, anyLogistix offers about 200 predefined KPIs. In each KPI group, we need to select the KPI and chart type (a table, line, bar chart, or histogram chart) (■ Fig. 5.7). We can also filter or detail KPI by products, types, and objects:

- Types: distribution center, factory, supplier, and customer

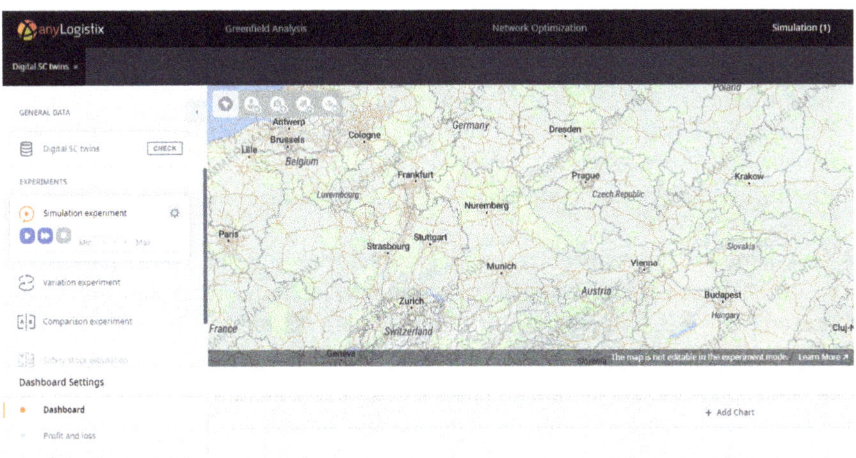


■ **Fig. 5.6** Adding a new KPI to a dashboard

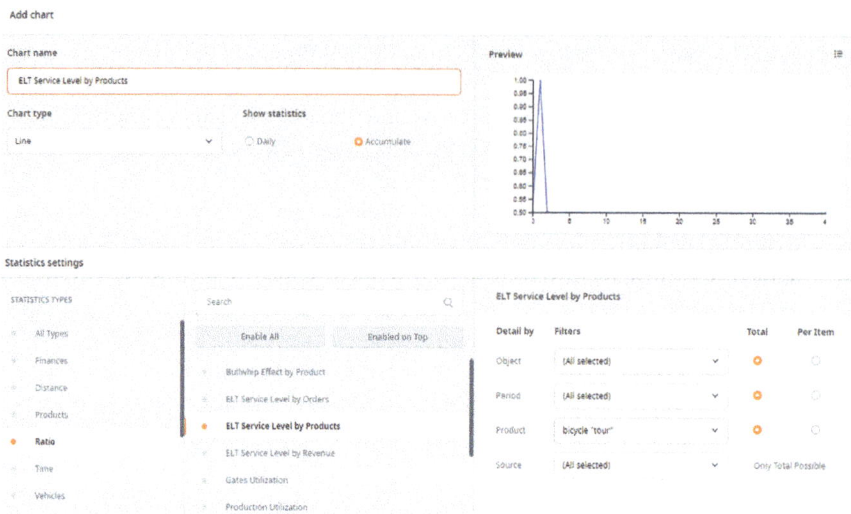


**Fig. 5.7** Selecting a KPI and its visualization form



- Objects: individual distribution centers, factories, suppliers, and customers
- Products: individual products

In Fig. 5.7, an example of selecting a KPI "Service level by products" with a visualization form "Line" and filtering for bicycle type "tour" is shown.

## 5.2.2 Simulation Model Design

In *simulation*, we extend our analysis from ▶ Chap. 4 (NO example) by adding the following features:

- We transit from flows (as in NO) to orders, i.e., the customer demand is no longer considered an aggregated flow during a period, but it is now generated as orders at certain intervals, e.g., 10 bicycles every day.
- We introduce inventory control to manage ordering processes.
- We introduce sourcing policies (e.g., single vs. multiple sourcing) to manage replenishment processes.
- We introduce shipment control (LTL/FTL) to manage shipment processes.

First, we simulate the supply chain with two factories in Poland and Germany and a DC in the Czech Republic subject to customer demand from ▶ Table 4.3 (see GFA exercise in ▶ Chap. 4) and the following data (Table 5.2).

For the DC and the factory, three inventory policies have to be developed. We assign the DC and factories a "min-max policy" for all products, where the minimum stock is 50, the maximum stock is 100, and the initial stock is 50 bicycles.

To evaluate the simulation results, we consider six KPIs according to the needs of Polarbear:

**Table 5.2** Parameters for simulation model

| Object | Inventory policy | | Expected lead time (ELT), days | Transportation time, days | Production time per unit, days | Sourcing policy | Transportation policy |
|---|---|---|---|---|---|---|---|
| | Min | Max | | | | | |
| Factory Germany | 50 | 100 | | 2 | 0.1 | | LTL |
| Factory Poland | 50 | 100 | | 2 | 0.07 | | LTL |
| DC Czech Republic | 50 | 100 | | 2 | | Closest dynamic | LTL |
| Customers | | | 5 | | | | |

1. Financial KPIs, such as profit, revenue, and costs
2. ELT service level by product, which is the ratio of products delivered within the expected lead time to the total ordered quantity
3. Demand fulfillment (product backlog)
4. Available inventory
5. Production capacity utilization
6. Lead time

With all of the parameters described, we now run the simulation for a period of 1 year.

**Creating an anyLogistix Model**

Step 1: Open scenario PB SIM Level 1.

Step 2: Check data from Table 5.2 in tables "DCs and Factories," "Inventory," "Production," "Sourcing," and "Unit Conversions." Explain the data in the aforementioned tables. The data in these tables should correspond to Table 5.2.

**Performing Experiments**

Step 1: Go to Simulation Experiment and run it.

Step 2: Analyze the results using the KPI dashboard "Revenue, Profit, Costs," "Lead Time," "ELT Service Level," "Production Utilization," "Demand (Product Backlog)," and "Average Available Inventory."

(a) What are the profit, revenue, and costs of the SC?
(b) Is demand for all customers satisfied? Explain.
(c) What is the production utilization of factories in Poland and Germany? Explain why the German factory is utilized to 100% and the factory in Poland is utilized

only to 75% even if we clearly see insufficient production quantities and the backlog?

(d) What is your judgment on the inventory dynamics in the SC?

(e) What is your judgment on the lead time?

(f) What suggestions for improvement do you have which could increase profit and customer satisfaction?

### ■ Solution

5

Step 1: Go to Simulation Experiment and run it (◘ Fig. 5.8).

Step 2: Analyze the results using the KPI dashboard "Revenue, Profit, Costs," "Lead Time," "ELT Service Level," "Production Utilization," "Demand (Product Backlog)," and "Average Available Inventory."

(a) What are the profit, revenue, and costs of the supply chain (◘ Table 5.3)?

(b) Is demand for all customers satisfied? Explain.

Not at all. We have a very high backlog and a decreasing service level.

(c) What is the production utilization of factories in Poland and Germany? Explain why the German factory is utilized to 100% and the factory in Poland is utilized only to 75% even if we clearly see insufficient production quantities and the backlog?

100% for the factory in Germany and 75% for the factory in Poland result from the constraints on production quantities and design capacity defined in tables "Inventory" and "Production," respectively (◘ Fig. 5.9).

Production time in table "Production" determines the *design capacity* of a factory (i.e., what is possible in principle). For example, a production time of

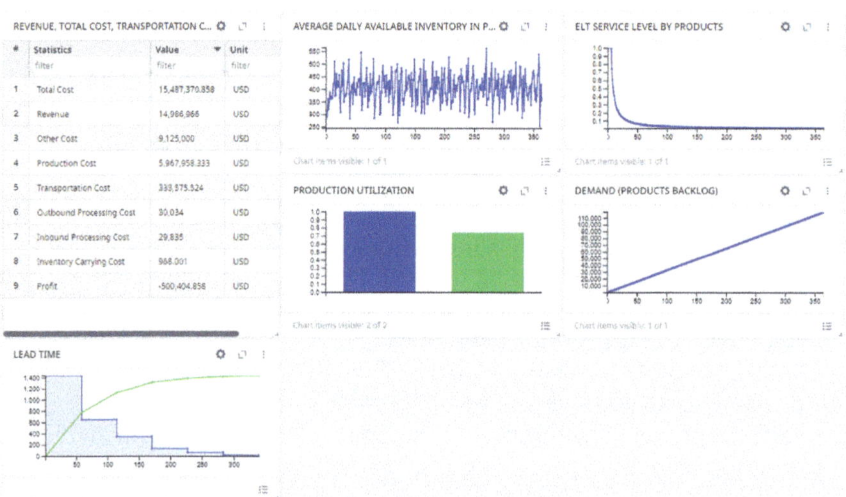


◘ **Fig. 5.8** Simulation results

**Table 5.3** Financial performance

| Statistics | Value (USD) |
|---|---|
| Total cost | 15,487,371 |
| Revenue | 14,986,966 |
| Other cost | 9,125,000 |
| Production cost | 5,967,958 |
| Transportation cost | 333,576 |
| Outbound processing cost | 30,034 |
| Inbound processing cost | 29,835 |
| Inventory carrying cost | 968 |
| Profit | −500,405 |

0.1 (day) means that we need 0.1 day to produce one bicycle, and so our daily maximum possible production rate is 10 bicycles. The *real production rate* is determined by the production control policy in table "Inventory" saying in our example that we should produce as many bicycles to satisfy the min = 50 and max = 100 constraints.

(d) What is your judgment on the inventory dynamics in the SC?
Fluctuations with a high amplitude are visible in the diagram.

(e) What is your judgment on the lead time?
It is very long and unequally distributed (e.g., high lead time fluctuations).

(f) What suggestions for improvement do you have which could increase profit and customer satisfaction?
Revenue is very low because customer demand is not fulfilled (see backlog and ELT service-level diagrams). Inventory control policy parameters (i.e., reorder point and target inventory) might need to be adjusted. Production capacity utilization is very high, which indicates a bottleneck. Capacity must be increased and aligned with demand, order frequency, and inventory control policy parameters.

Polarbear's supply chain managers suggest the following changes to the input parameters (Table 5.4).

Having changed all the parameters, we now run new simulation.

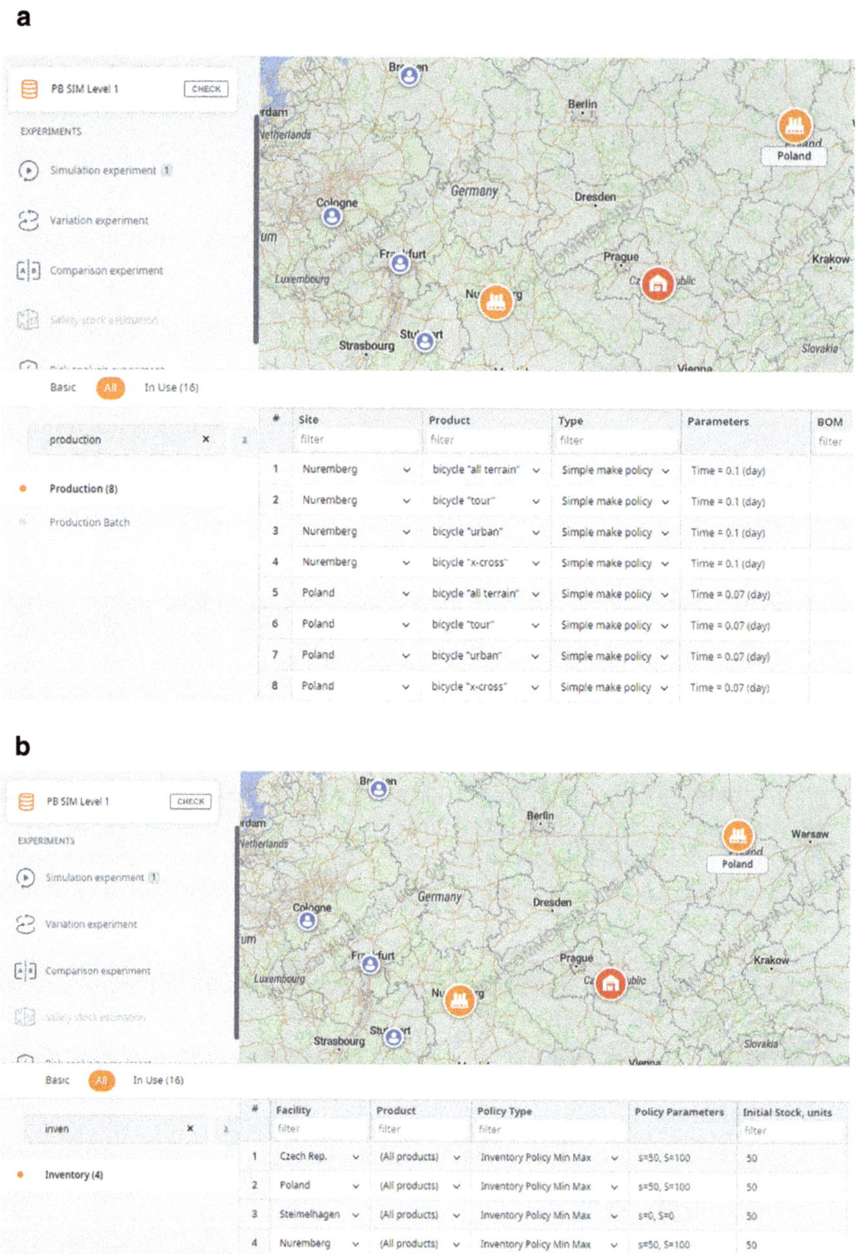


**Fig. 5.9** Production and inventory control policies

## ▪ Creating an anyLogistix Model

Step 1: Open scenario PB SIM Level 1_Improved.

Step 2: Check data from Table 5.4 in tables "Inventory" and "Production." The data in these tables should correspond to Table 5.4.

**Table 5.4** Adjusted parameters for simulation model

| Object | Inventory policy | | ELT, days | Transportation time, days | Production time per unit, days | Sourcing policy | Transportation policy |
|---|---|---|---|---|---|---|---|
| | Min | Max | | | | | |
| Factory Germany | | | 5 | 2 | | | LTL |
| *X-cross* | 30 | 60 | | | 0.05 | | |
| *Urban* | 100 | 200 | | | 0.015 | | |
| *All terrain* | 40 | 80 | | | 0.04 | | |
| *Tour* | 75 | 150 | | | 0.02 | | |
| Factory Poland | | | 5 | 2 | | | LTL |
| *X-cross* | 30 | 60 | | | 0.05 | | |
| *Urban* | 100 | 200 | | | 0.015 | | |
| *All terrain* | 40 | 80 | | | 0.04 | | |
| *Tour* | 75 | 150 | | | 0.02 | | |
| DC Czech Republic | | | | 2 | | Closest dynamic | LTL |
| *X-cross* | 60 | 120 | | | | | |
| *Urban* | 200 | 400 | | | | | |
| *All terrain* | 80 | 160 | | | | | |
| *Tour* | 150 | 300 | | | | | |
| Customers | | | 5 | | | Closest dynamic | |

## Performing Experiments

Step 1: Go to Simulation Experiment and run it.

Step 2: Analyze the results using the KPI dashboard "Revenue, Profit, Costs," "Lead Time," "ELT Service Level," "Production Utilization," "Demand (Product Backlog)," and "Average Available Inventory."

(a) What are the profit, revenue, and costs of the SC? Did we improve?
(b) Is demand for all customers satisfied? Explain.
(c) What is the production utilization of factories in Poland and Germany? Explain.
(d) What is your judgment on the inventory dynamics in the SC?
(e) Explain how MIN (reorder point) and MAX (target inventory) values have been computed.
(f) What is your judgment on the lead time?
(g) Explain why the changes made improved supply chain performance.
(h) How can you validate the simulation modeling results using the previous network optimization experiments?

- **Solution**

Step 1: Go to Simulation Experiment and run it (◘ Fig. 5.10).

Step 2: Analyze the results using the KPI dashboard "Revenue, Profit, Costs," "Lead Time," "ELT Service Level," "Production Utilization," "Demand (Products Backlog)," and "Average Available Inventory."

(a) What are the profit, revenue, and costs of the supply chain? Did we improve? (◘ Table. 5.5).

Yes, we significantly improved revenues and profits.

(b) Is demand for all customers satisfied? Explain.

Yes. The service level is 100%, and there is no backlog.

(c) What is the production utilization of factories in Poland and Germany? Explain.

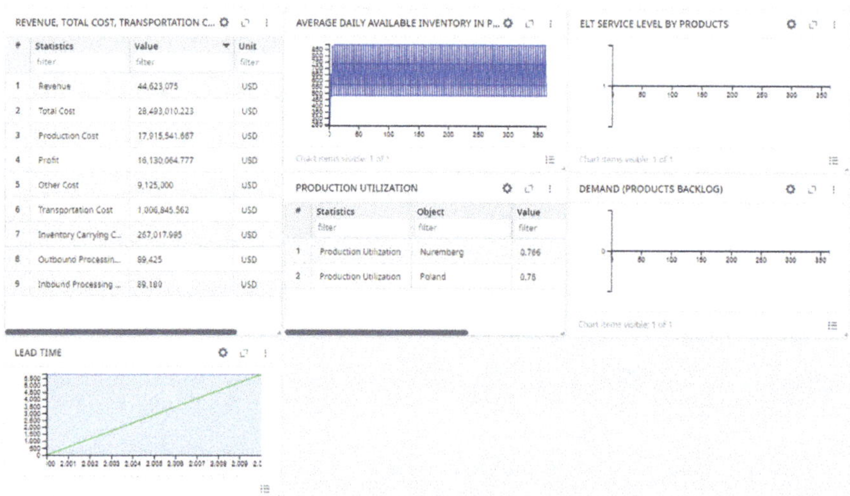


◘ **Fig. 5.10** Simulation results after improvement

◘ **Table 5.5** Financial performance after improvement

| Statistics | Value (USD) |
|---|---|
| Revenue | 44,623,075 |
| Inventory carrying cost | 267,018 |
| Inbound processing cost | 89,180 |
| Other cost | 9,125,000 |
| Outbound processing cost | 89,425 |
| Production cost | 17,915,542 |
| Transportation cost | 1,006,846 |
| Total cost | 28,493,010 |
| Profit | 16,130,065 |

78%. There is no longer production backlog in the supply chain.

(d) What is your judgment on the inventory dynamics in the supply chain?
Stable inventory dynamics.

(e) Explain how MIN (reorder point) and MAX (target inventory) values were computed.
$s$ = demand per day * transportation time; $S = 2*s$

(f) What is your judgment on the lead time?
It became constant at a value of 2 days because there was no backlog or product shortage.

(g) Explain why the changes made improved supply chain performance.
The adjustment of inventory policy parameters along with increasing production capacity and aligning it with demand, ordering frequency, and inventory policy improved the performance of the supply chain.

(h) How can you validate the simulation modeling results using the network optimization (NO) experiments?
Profit in the simulation experiment is close to the profit obtained in the network optimization for the same supply chain design and experimental dataset.

## 5.2.3 Disruption Simulation

We are now interested to perform a *stress test* of our supply chain. In particular, what happens if the DC would be disrupted for 2 months?

### ▪ Creating an anyLogistix Model

Step 1: Create a copy of your scenario PB SIM Advanced_Improved and rename PB SIM Advanced_Improved_Risk.

Step 2: Use table "Event" to set up a breakdown of the DC for 2 months starting July 1.

### ▪ Performing Experiments

Step 1: Go to Simulation Experiment and run it.

Step 2: Analyze the results using the KPI dashboard "Revenue, Profit, Costs," "Lead Time," "ELT Service Level/Service Level," "Production Utilization," "Demand (Product Backlog)," and "Average Available Inventory."

(a) How does a disruption of the DC impact the KPIs?
(b) How would you suggest managing disruption risks in the supply chain?

### ▪ Solution

*Creating an anyLogistix Model*

Step 1: Create a copy of your scenario PB SIM Advanced_Improved and rename PB SIM Advanced_Improved_Risk.

Step 2: Use table "Event" to set up a breakdown of the DC for 2 months starting July 1 (◘ Fig. 5.11).

*Performing Experiments*

Step 1: Go to Simulation Experiment and run it (◘ Fig. 5.12).

Step 2: Analyze the results using the KPI dashboard "Revenue, Profit, Costs," "Lead Time," "ELT Service Level/Service Level," "Production Utilization," "Demand (Product Backlog)," and "Average Available Inventory."

(a) How does a disruption of the DC impact the KPIs?

Service level, inventory, and backlog diagrams demonstrate the impact of a DC disruption from July 1 to August 31. Service level and backlog decrease in this period, and there are no deliveries to customers (see inventory diagram). In total, there are also decreases in revenue and profit if we compare simulations with and without disruptions (◘ Table. 5.6).

(b) How would you suggest managing disruption risks in the supply chain?

5

The DC in Germany could be used as a backup in case of disruptions. Alternatively, a smaller regional warehouse downstream in the supply chain between the main DC and customers could be installed. This would ensure

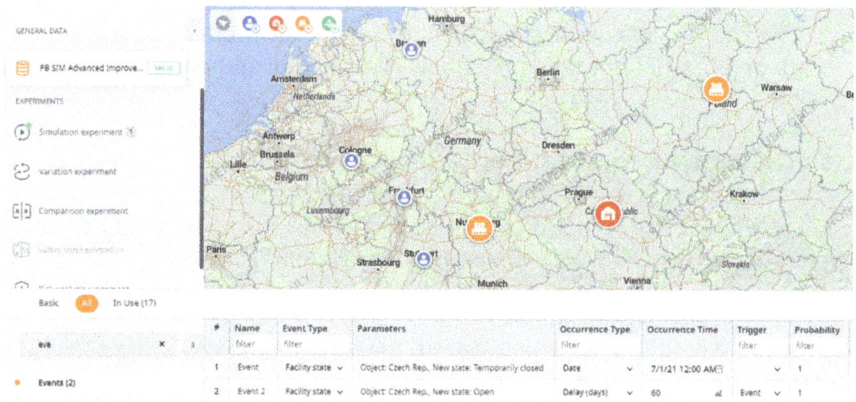


◘ **Fig. 5.11** Disruption scenario definition

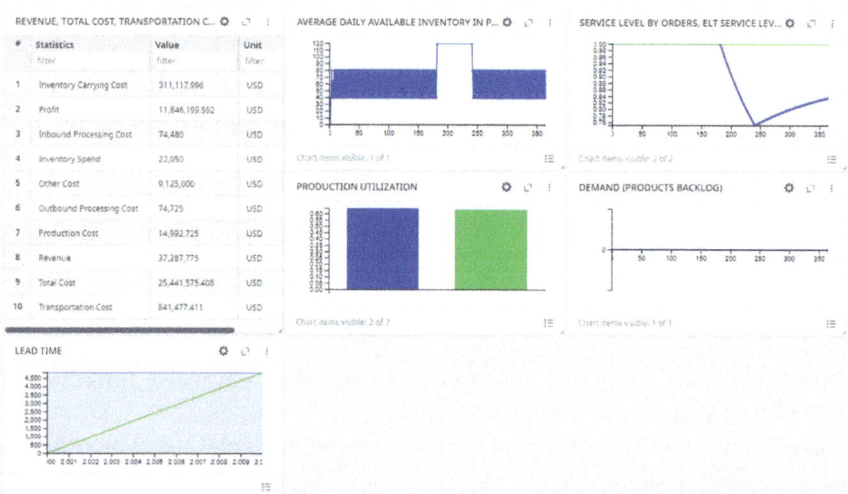


◘ **Fig. 5.12** Resilience simulation results

**Table 5.6** Comparison of financial performance without and with disruptions

| Statistics | Without disruptions | With disruptions |
|---|---|---|
| Revenue | 44,623,075 | 37,287,775 |
| Inventory carrying cost | 267,018 | 311,118 |
| Inbound processing cost | 89,180 | 74,480 |
| Inventory spend | – | 22,050 |
| Other cost | 9,125,000 | 9,125,000 |
| Outbound processing cost | 89,425 | 74,725 |
| Production cost | 17,915,542 | 14,992,725 |
| Transportation cost | 1,006,846 | 841,477 |
| Total cost | 28,493,010 | 25,441,575 |
| Profit | 16,130,065 | 11,846,200 |

that some inventory was in place for risk mitigation, ensure deliveries to customers in the event of disruption upstream, and prevent the ripple effect in the supply chain.

## 5.3 anyLogistix Case Study: Comparison and Variation Experiments

### 5.3.1 Comparison Experiment

In order to simplify a comparison of three considered simulation scenarios, we run a comparison experiment.

- **Creating a Comparison Experiment**

Step 1: Open scenario PB SIM Advanced and go to Comparison Experiment.

Step 2: Select scenarios PB SIM Advanced _Improved, PB SIM Advanced _ Improved_Risk, and PB SIM Advanced to compare.

Step 3: Create KPI dashboard. The KPIs for a comparative analysis of three scenarios should be "Demand (backlog)," "Profit," "ELT Service Level by Orders," and "Service Level by Orders."

- **Performing Experiments**

Step 1: Run Comparison experiment

Step 2: Analyze the results.

(a) What are the profit, ELT service level, service level, and order backlog for the two scenarios? Did we improve in the Improved scenario?
(b) How can you explain the relationship between the Simulation and Comparison experiments?

- **Solution**

Step 1: Open scenario PB SIM Advanced and go to Comparison Experiment.

Step 2: Select scenarios PB SIM Advanced _Improved, PB SIM Advanced _ Improved_Risk, and PB SIM Advanced to compare.

Step 3: Create KPI dashboard (◘ Fig. 5.13).

Note that in order to create the KPI dashboard in the Comparison experiment, we should first go to Configure Statistics, select the KPIs we want to add, and then add them to the dashboard through Add Item as explained in ► Sect. 5.2.1.

- **Performing Experiments**

Step 1: Run Comparison experiment.

Step 2: Analyze the results (◘ Table. 5.7).

(a) What is the profit, ELT service level, service level, and order backlog in three scenarios? Did we improve in the second scenario?

Yes, we improved all KPIs.

(b) How can you explain the relationship between the Simulation and Comparison experiments?

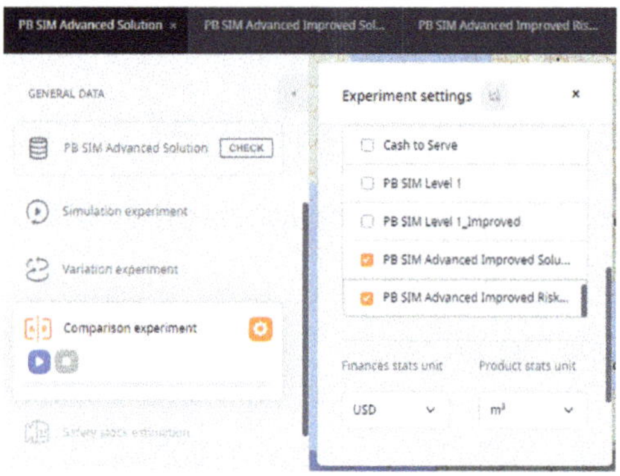


◘ **Fig. 5.13** Comparison experiment setup

◘ **Table 5.7** Comparison experiment results

| | Service level by orders | Demand (orders backlog) | Profit | ELT service level by products | Total cost |
|---|---|---|---|---|---|
| PB SIM advanced solution | 0.006 | 3866 | −500,404.9 | 0.008 | 15,487,370.9 |
| PB SIM advanced improved solution | 1 | 0 | 16,130,064.8 | 1 | 28,493,010.2 |
| PB SIM advanced improved risk solution | 0.836 | 0 | 11,846,199.6 | 1 | 25,441,575.4 |

The Comparison experiment simultaneously runs multiple simulation scenarios. When only the overall KPIs are important for decision-making support, this is much faster than running multiple simulations separately. However, in a Comparison experiment, the process and dynamics of each scenario cannot be observed.

### 5.3.2 Variation Analysis

Rather than running the same simulation multiple times with different parameter values or combinations, the variation experiment allows multiple variations of the same simulation to be run simultaneously. A variation experiment highlights how KPIs change depending on different parameter values. This kind of *sensitivity analysis* can also be used to verify the validity of the results of the simulation model.

A variation analysis must be performed to check Polarbear's simulation model and the results. The daily demand for "urban" bicycles in Cologne can be varied with a minimum demand of 50 and a maximum demand of 170 in steps of 40. The variation should be performed for a period of 1 year.

#### ■ Creating a Variation Experiment

Step 1: Open scenario PB SIM Advanced_Improved and go to Variation Experiment.

Step 2: Select the parameter we will vary (Inventory—Min).

Step 3: Create KPI dashboard. The KPIs for a variation analysis should be "Demand (backlog)," "Profit," and "ELT Service Level."

#### ■ Performing Experiments

Step 1: Run Variation experiment.

Step 2: Analyze the results using the KPIs "Demand (backlog)," "Profit," and "ELT Service Level."

(a) What is the profit, ELT service level, and order backlog for different demands?
(b) Why and how do the KPIs change as demand changes?
(c) Is the simulation model sensitive?

#### ■ Solution

*Creating a Variation Experiment*

Step 1: Open scenario Advanced_Improved and go to Variation Experiment.

Step 2: Select the parameter we will vary (Inventory—Min).

Step 3: Create KPI dashboard. The KPIs for a comparative analysis of the two scenarios should be "Demand (backlog)," "Profit," and "ELT Service Level" (◘ Fig. 5.14).

*Performing Experiments*

Step 1: Run Variation experiment.

Step 2: Analyze the results using the KPIs "Demand (backlog)," "Profit," and "ELT Service Level."

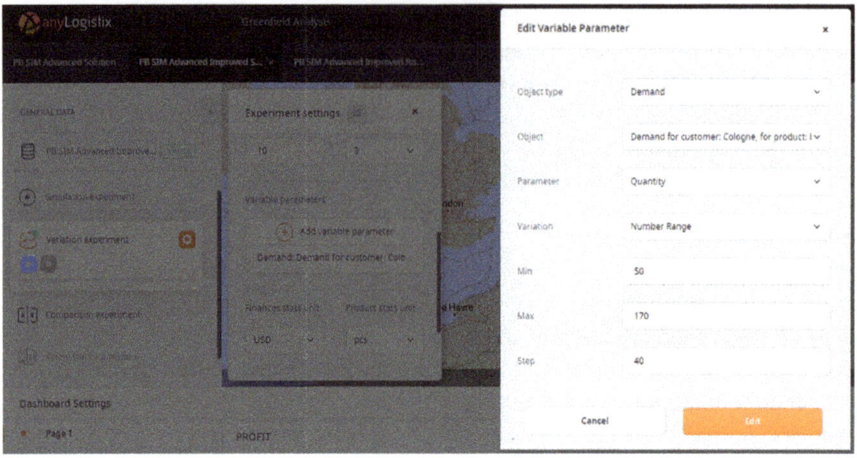


**Fig. 5.14** Variation experiment setup

**Table 5.8** Variation experiment results

| | Demand (orders backlog) | Profit | ELT service level by products | Total cost |
|---|---|---|---|---|
| Quantity: 50.0 | 0 | 16,130,064.8 | 1 | 28,493,010.2 |
| Quantity: 90.0 | 858 | 11,641,000.3 | 0.516 | 26,984,094.7 |
| Quantity: 130.0 | 1028 | 11,658,019.7 | 0.452 | 26,987,035.3 |
| Quantity: 170.0 | 1272 | 11,587,565.8 | 0.402 | 26,987,629.2 |

(a) What is the profit, ELT service level, and order backlog for different demands? (Table. 5.8).

(b) Why and how do the KPIs change as demand changes?

Our supply chain is not well prepared for an increase in demand in Cologne. Due to limited production capacity and selected reorder points and target inventories, the demand increase in Cologne would result in a backlog and significant decreases in service level and profit.

(c) Is the simulation model sensitive?

Yes, it is. Changes in demand have a direct impact on service level, profit, and backlog.

## 5.3.3 Managerial Recommendation

Develop recommendations for Polarbear Bicycle's management. Consider GFA, NO, SIM, comparison, and variation results. What supply chain design would you recommend? Consider the impact of inventory control and production capacities.

Are there any other factors that should be taken besides the results of the experiments performed?

**Solution**

Given the results of all the analyses, several recommendations can be made concerning how Polarbear Bicycle can improve their SC. First, managing the inventory is key to keeping costs low, and thereby increasing profit. By varying inventory control policy parameters, we could achieve significant improvements in profit and service level.

Second, if net profit is the most important KPI for Polarbear, the results of the NO indicate that the best network design is to take the opportunity to rent the factory in Poland and the DC in the Czech Republic while maintaining production capacity in Nuremberg. Along the same lines, this would also mean not opening the DC in Germany. However, other factors should be taken into account concerning this option since the closing costs for the DC in Germany were not considered in the software model. In addition, a second DC could be useful as backup in case severe disruptions occur in the SC.

Third, the variation analysis showed that profit and service level are highly sensitive to demand changes. As such, capacity flexibility in production and adjustment of inventory control policies should be considered in order to cope with unexpected demand fluctuations.

## 5.4 Digital Twin-Based Supply Chain Analytics

### 5.4.1 Digital Supply Chain Twin

Supply chains are evolving towards technology-driven networks and digital ecosystems (Ivanov et al., 2021b). Data analytics, additive manufacturing, and Industry 4.0 allow creating end-to-end visibility for supply chains based on dynamically reconfigurable material flows and digital information flows. For example, digital supply chain financing with the use of blockchain (e.g., deep-tier financing) creates an end-to-end supply chain visibility which is of vital importance for supply chain viability and resilience.

Digital technology for supply chain resilience management can be classified into

- Visualization
- Early warning systems
- Blockchain and supply chain financing
- Real-time disruption detection systems

Data-driven disruption modeling provides a basis for proactive, resilient supply chain design in anticipation of disruptions and structural-parametrical adaptation in the event of disruptions. The modeling combines simulation, optimization, and data analytics to create a *digital supply chain twin* and thereby manage disruption risks.

A *digital twin* of a supply chain is "a virtual system comprised of (1) a digital visualization of a physical supply chain and its elements (e.g., firms, flows, and products) in a computer model, (2) digital technologies providing data about the physical object (e.g., sensors, blockchain, clouds), and (3) descriptive, predictive, and prescriptive analytics for decision-making support" (Ivanov, 2023a).



Tozanli and Saénz (2022) define a digital supply chain twin as "a combination of multiple enabling technologies, such as sensors, cloud computing, AI [artificial intelligence] and advanced analytics, simulation, visualization, and augmented and virtual reality." Dolgui and Ivanov (2023) consider a *digital supply chain twin* a computerized model that at each point of time represents the physical supply chain with the actual transportation, inventory, demand, and capacity data and can be used for planning and real-time control decisions. In ◘ Fig. 5.15, we illustrate the concept and design of digital supply chain twins.

A digital twin is linked to a real object. Through continuous data exchange, the digital twin always behaves synchronously with its real counterpart, and they influence each other. It mirrors both the state and the performance of the real system and can also send control signals back to the real system. Digital twins can be used for descriptive (e.g., performance visualization), predictive (e.g., demand forecasting), and prescriptive (e.g., supply chain recovery simulation and optimization) decision-making support (Burgos & Ivanov, 2021; Rolf et al., 2022).

❯ While simulation is a major part of digital twins, digital twins differ from standalone simulation models in several aspects (Ivanov, 2023b). First, a digital twin is a combination of different models (e.g., optimization, simulation, and visualization). Second, a digital twin includes not only internal data processing for modeling, but also pre- and post-processing of data in integration with external systems. Third, a digital twin is a knowledge-driven and knowledge-transferring system using human-machine interfaces. Fourth, a digital twin is supported by a technology, which implements data collection, processing, visualization, and interoperability.

In the digital supply chain twin, *model-based decision-making support* enables simulation of the supply chain's dynamic behavior in the event of disruption (Ivanov & Dolgui, 2021b). In addition, before a disruption occurs, potential impacts on supply chain performance can be evaluated, and then recovery policies can be optimized. *Data analytics* is used at the proactive stage for building realistic disruption scenarios based on risk data about historical disruptions and other data (e.g., supplier reliability data from ERP systems) during the supply chain design phase. At the reactive stage, data analytics is used for disruption identification in real time using process feedback data, e.g., from sensors, T&T, and RFID.

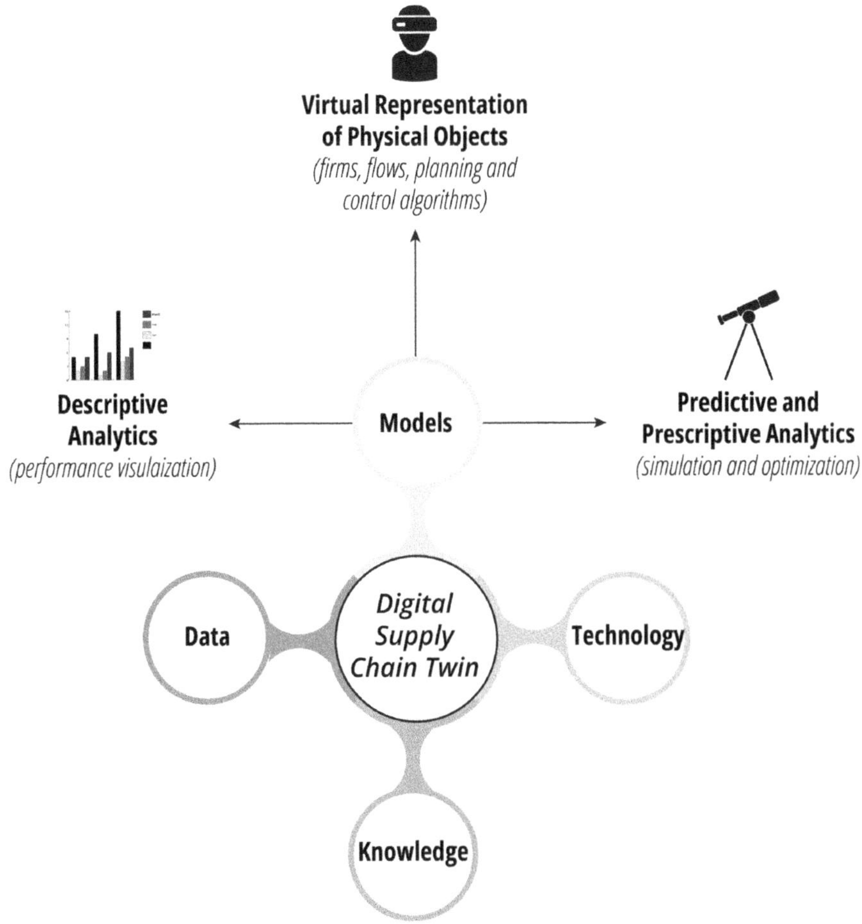


**Fig. 5.15** Concept and design of a digital supply chain twin (based on Ivanov, 2023a, 2023b)

## 5.4.2 anyLogistix as a Digital Supply Chain Twin

With the help of anyLogistix supply chain simulation and optimization software, a digital supply chain can be designed. Figure 5.16 shows the structure of a digital supply chain twin created in this study for disruption analysis.

The digital supply chain twin encompasses three major perspectives—the network, the flows, and the parameters. The supply chain network can be designed using different location objects, such as customers, DCs, factories, and suppliers. The flows in the network can be flexibly arranged to represent the specifics of different supply chains. The flows are associated with some design (i.e., maximum) capacities in production, warehouses, and transportation and controlled by the associated production, inventory, sourcing, and shipment policies. These policies

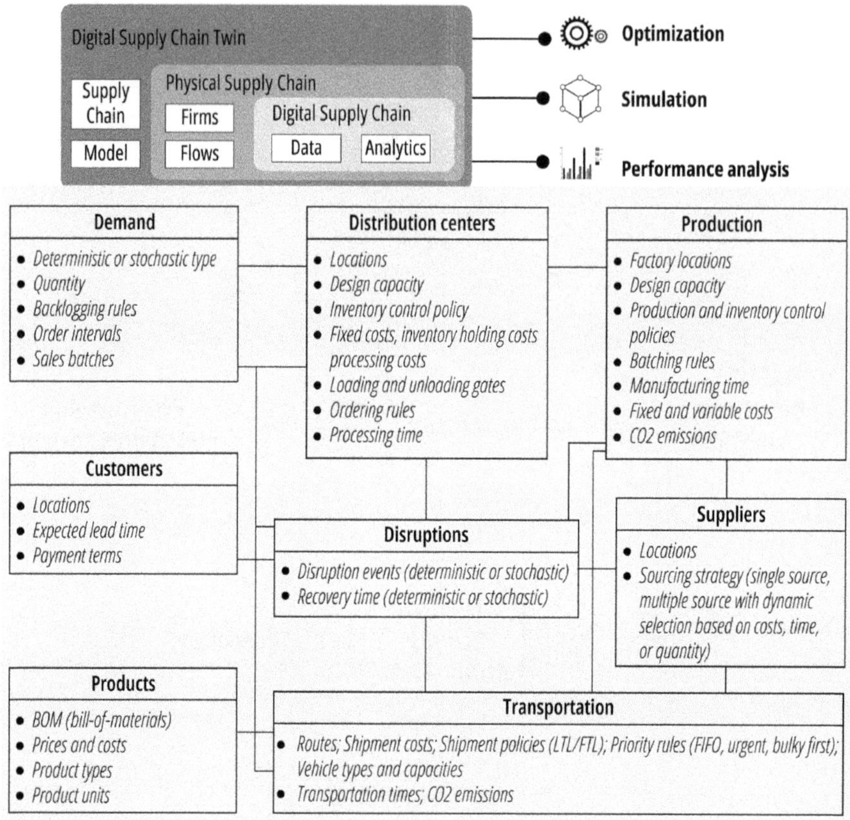


**Fig. 5.16** Digital supply chain design for disruption analysis using anyLogistix (based on Burgos & Ivanov, 2021)

can be flexibly adapted to the specifics of the supply chain and its management rules. Finally, different operational parameters such as demand, lead time, and control policies' thresholds (e.g., reorder point, target inventory, and minimum vehicle load) can be defined. With that functionality, a digital model of a physical supply chain (i.e., a digital supply chain) can be created and used for optimization and simulations to analyze supply chain operations and performance dynamics under disruptions.

The following modeling framework for resilient supply chain design and recovery planning, which contains the conceptually integrated data, can be proposed (Fig. 5.17).

Figure 5.17 depicts data sources, information systems, and parameters of models for resilient supply chain design and recovery planning where the respective data is used. We illustrate the digital supply chain twin principles and data-driven disruption modeling framework from Fig. 5.17 for an example of a digital supply chain twin developed with the use of anyLogistix software, which is a tool for model-based decision-making support in supply chains (Fig. 5.18).

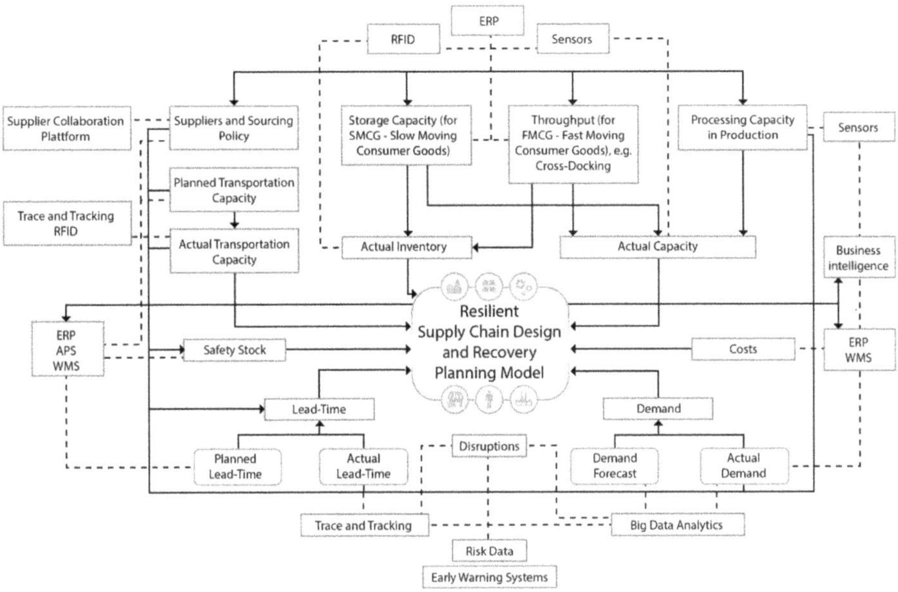


**Fig. 5.17** Data structure in supply chain disruption risk modeling framework (Ivanov & Dolgui, 2021b)

The combination of anyLogistix and data analytics is based on a mapping of the risk data with geographical locations in the supply chain structure as discussed in Ivanov et al. (2019a, 2019b). This mapping is considered at the resilient supply chain design and resilience recovery stages. The decision support system (DSS) uses disruption risk data to assess supplier and transportation disruption risks, predicting possible supply chain interruptions. This data is used for computing of alternative supply network topologies and backup routes with assessment of estimated times of arrival in anyLogistix.

In the dynamic mode, simulation is applied using real-time data to analyze the disruption impact on supply chain performance and alternative supply chain designs that contain non-disrupted network nodes and arcs depending on real-time inventory, demand, and capacity data. Furthermore, the interaction of data analytics and simulation-optimization tools is not limited to updating model data. Considering the output of simulation modeling, simulation results can be transferred to an ERP system or a business intelligence (BI) tool in order to analyze the performance impact of disruptions. Additionally, the simulation models can activate some BI algorithms. For example, if service level decreases to a certain level in the supply chain's simulation model, the digital twin might activate a BI algorithm to search for the cause of that problem and the necessary data updated to resolve the problem.

Figure 5.19 demonstrates three major areas of supply chain disruption risk management covered in the digital twin proposed, i.e., disruption identification, disruption modeling, and disruption impact assessment.

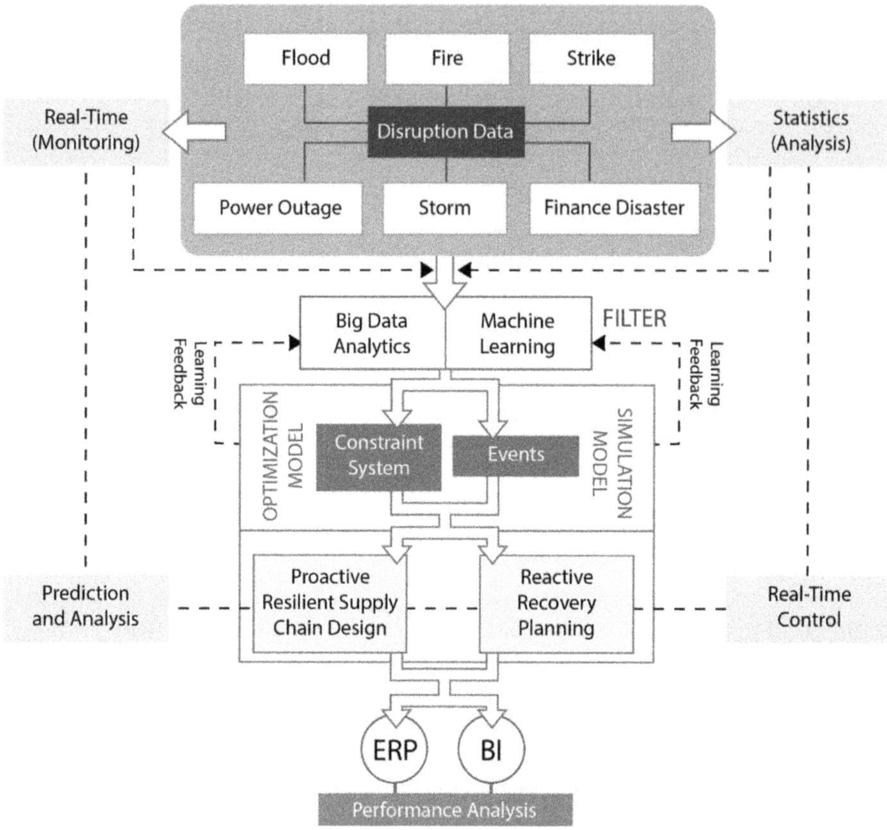


**Fig. 5.18** Digital supply chain twin for managing disruption risks

Figure 5.19 shows the mapping of risk data identification, locational supply chain model, and performance impact analysis. At the analysis stage, historical risk data from external databases (e.g., natural disaster events in the past and geographical regional risk assessments) and internal sources (e.g., ERP data about supplier reliability performance) are collected. This data can be used to build disruption scenarios for supply chain resilience analysis, which then can be examined and tested in simulation-optimization models. In the real-time mode, the resilience analytics system is used to search for relevant disruption data that might affect supply chain resilience. The simulation can be subsequently run to observe the impact of such a disruption on supply chain performance. Moreover, some recovery policies, such as activation of alternative supply chain designs during the disruption, are simulated. Data collected from different sources in real time are used to update model parameters, such as capacity, inventory, and lead times in terms of production and shipment capacities and inventory availability in the supply chain. Finally, data analytics can be used as a data-driven *learning* system to use past experiences to manage future disruptions and so utilizing cyber-physical, artificial intelligence, and machine learning principles and technologies (Ivanov (2024).

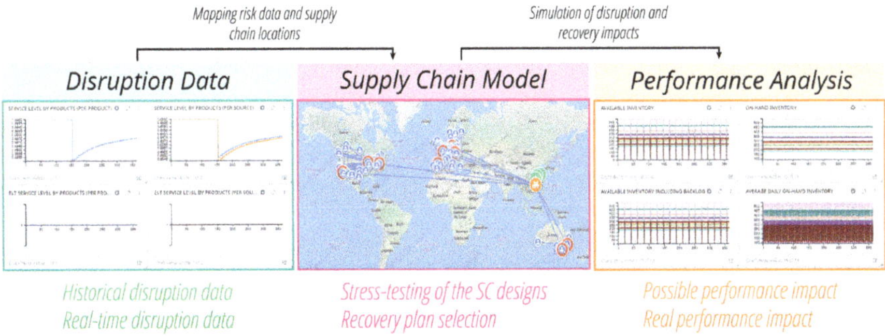


**Fig. 5.19** Interrelations between risk data, modeling, and performance analysis (based on Ivanov & Dolgui, 2021b)

## References


Aldrighetti, R., Battini, D., & Ivanov, D. (2023). Efficient resilience portfolio design in the supply chain with consideration of preparedness and recovery investments. *Omega, 117*, 103841.

Basole, R. C., & Bellamy, M. A. (2014). Supply network structure, visibility, and risk diffusion: A computational approach. *Decision Sciences, 45*(4), 1–49.

Burgos, D., & Ivanov, D. (2021). Food retail supply chain resilience and the COVID-19 pandemic: A digital twin-based impact analysis and improvement directions. *Transportation Research—Part E: Logistics and Transportation Review, 152*, 102412.

Chopra, S., & Sodhi, M. S. (2004). Managing risk to avoid supply-chain breakdown. *MIT Sloan Management Review, 46*, 52–61.

Chopra, S., Reinhardt, G., & Mohan, U. (2007). The importance of decoupling recurrent and disruption risks in a supply chain. *Naval Research Logistics, 54*(5), 44–555.

Christopher, M. (2011). *Logistics & supply chain management* (4th ed.). Financial Times Prentice Hall.

Dolgui, A., Ivanov, D. (2023). Metaverse supply chain and operations management. *International Journal of Production Research, 61*(23), 8179–8191. https://doi.org/10.1080/00207543.2023.2240900

Dolgui, A., Ivanov, D., & Sokolov, B. (2018). Ripple effect in the supply chain: An analysis and recent literature. *International Journal of Production Research, 56*(1–2), 414–430.

Dolgui, A., Ivanov, D., & Rozhkov, M. (2020). Does the ripple effect influence the bullwhip effect? An integrated analysis of structural and operational dynamics in the supply chain. *International Journal of Production Research, 58*(5), 1285–1301.

Dolgui, A., Gusikhin, O., Ivanov, D., Li, X., & Stecke, K. (2024). A network-of-networks adaptation for cross-industry manufacturing repurposing. *IISE Transactions*. *56*(6), 666–682. https://doi.org/10.1080/24725854.2023.2253881.

Hägele, S., Grosse, E., & Ivanov, D. (2023). Supply chain resilience: A tertiary study. *International Journal of Integrated Supply Management, 16*(1), 52–81.

Hosseini, S., Ivanov, D., & Dolgui, A. (2019). Review of quantitative methods for supply chain resilience analysis. *Transportation Research: Part E, 125*, 285–307.

Ivanov, D. (2018). *Structural dynamics and resilience in supply chain risk management*. Springer.

Ivanov, D. (2021a). *Introduction to supply chain resilience*. Springer Nature.

Ivanov, D. (2021b). Supply chain viability and the COVID-19 pandemic: A conceptual and formal generalisation of four major adaptation strategies. *International Journal of Production Research, 59*(12), 3535–3552.

Ivanov, D. (2022a). Viable supply chain model: Integrating agility, resilience and sustainability perspectives—Lessons from and thinking beyond the COVID-19 pandemic. *Annals of Operations Research, 319*, 1411–1431.

Ivanov, D. (2022b). Lean resilience: AURA (active usage of resilience assets) framework for post-COVID-19 supply chain management. *International Journal of Logistics Management, 33(4)*, 1196–1217.

Ivanov, D. (2023a). Conceptualisation of a 7-element digital twin framework in supply chain and operations management. *International Journal of Production Research, 62(6)*, 2220–2232. https://doi.org/10.1080/00207543.2023.2217291

Ivanov, D. (2023b). Intelligent digital twin (iDT) for supply chain stress-testing, resilience, and viability. *International Journal of Production Economics, 263*, 108938.

Ivanov, D. (2023c). The Industry 5.0 framework: Viability-based integration of the resilience, sustainability, and human-centricity perspectives. *International Journal of Production Research, 61*(5), 1683–1695.

Ivanov, D. (2023d). Two views of supply chain resilience. *International Journal of Production Research*. https://doi.org/10.1080/00207543.2023.2253328


5

Ivanov, D. (2024). Conceptualisation of a 7-element digital twin framework in supply chain and operations management. *International Journal of Production Research, 62*(6), 2220–2232.

Ivanov, D., & Dolgui, A. (2019). Low-certainty-need (LCN) supply chains: A new perspective in managing disruption risks and resilience. *International Journal of Production Research, 57*(15–16), 5119–5136.

Ivanov, D., & Dolgui, A. (2020). Viability of intertwined supply networks: Extending the supply chain resilience angles towards survivability. A position paper motivated by COVID-19 outbreak. *International Journal of Production Research, 58*(10), 2904–2915.

Ivanov, D., & Dolgui, A. (2021a). OR-methods for coping with the ripple effect in supply chains during COVID-19 pandemic: Managerial insights and research implications. *International Journal of Production Economics, 232*, 107921.

Ivanov, D., & Dolgui, A. (2021b). A digital supply chain twin for managing the disruptions risks and resilience in the era of Industry 4.0. *Production Planning and Control, 32*(9), 775–788.

Ivanov, D., & Keskin, B. (2023). Post-pandemic adaptation and development of supply chain viability theory. *Omega, 116*, 102806.

Ivanov, D., Sokolov, B., & Dolgui, A. (2014). The ripple effect in supply chains: Trade-off 'efficiency-flexibility-resilience' in disruption management. *International Journal of Production Research, 52*(7), 2154–2172.

Ivanov, D., Dolgui, A., & Sokolov, B. (Eds.). (2019a). *Handbook of ripple effects in the supply chain*. Springer.

Ivanov, D., Dolgui, A., & Sokolov, B. (2019b). The impact of digital technology and Industry 4.0 on the ripple effect and supply chain risk analytics. *International Journal of Production Research, 57*(3), 829–846.

Ivanov, D., Tsipoulanidis, A., & Schönberger, J. (2021a). *Global supply chain and operations management: A decision-oriented introduction into the creation of value* (3rd ed.). Springer Nature. ISBN: 978-3-030-72331-6.

Ivanov, D., Tang, C. S., Dolgui, A., Battini, D., & Das, A. (2021b). Researchers' perspectives on Industry 4.0: Multi-disciplinary analysis and opportunities for operations management. *International Journal of Production Research, 59*(7), 2055–2078.

Ivanov, D., Dolgui, A., Blackhurst, J., & Choi, T.-M. (2023a). Toward supply chain viability theory: From lessons learned through COVID-19 pandemic to viable ecosystems. *International Journal of Production Research, 61*(8), 2402–2415.

Ivanov, D., Dolgui, A., Blackhurst, J., & Choi, T. M. (2023b). Viability of supply networks and ecosystems: Lessons learned from COVID-19 outbreak. *International Journal of Production Research, 61*(8), 2402–2415.

Ivanov D. (2024). Transformation of supply chain resilience research through the COVID-19 pandemic. *International Journal of Production Research*, https://doi.org/10.1080/00207543.2024.2334420

Ivanov D. (2024). Supply Chain Resilience: Conceptual and Formal Models Drawing from Immune System Analogy. *Omega, 127*, 103081.

Jackson I., Ivanov D. (2023). A beautiful shock? Exploring the impact of pandemic shocks on the accuracy of AI forecasting in the beauty care industry. *Transportation Research Part E: Logistics and Transportation Review* 180, 103360.

Jackson, I., Ivanov, D., Dolgui A., & J. Namdar (2024) Generative artificial intelligence in supply chain and operations management: a capability-based framework for analysis and implementation, *International Journal of Production Research*, https://doi.org/10.1080/00207543.2024.2309309.
Li, Y., Zobel, C. W., Seref, O., & Chatfield, D. C. (2020). Network characteristics and supply chain resilience under conditions of risk propagation. *International Journal of Production Economics, 223*, 107529.
Li, Y., Chen, K., Collignon, S., & Ivanov, D. (2021). Ripple effect in the supply chain network: Forward and backward disruption propagation, network health and firm vulnerability. *European Journal of Operational Research, 291*(3), 1117–1131.
MacCarthy, B., Ahmed, W., & Demirel, G. (2022). Mapping the supply chain: Why, what and how? *International Journal of Production Economics, 250*, 108688.
Rolf, B., Jackson, I., Müller, M., Lang, S., Reggelin, T., & Ivanov, D. (2022). A review on reinforcement learning algorithms and applications in supply chain management. *International Journal of Production Research. 61(20)*, 7151–7179.
Sawik, T. (2020). *Supply chain disruption management* (2nd ed.). Springer.
Tang, C. (2006). Perspectives in supply chain risk management. *International Journal of Production Economics, 103*(2), 451–488.
Tozanli, Ö., & Saénz, M. E. (2022). Unlocking the potential of digital twins in supply chains. *MIT Sloan Management Review, 63*, 1–4.

## Further Reading

Ambulkar, S., Blackhurst, J., & Grawe, S. (2015). Firm's resilience to supply chain disruptions: Scale development and empirical examination. *Journal of Operations Management, 33*, 111–122.
Blackhurst, J., Craighead, C. W., Elkins, D., & Handfield, R. (2005). An empirically derived agenda of critical research issues for managing supply-chain disruptions. *International Journal of Production Research, 43*(19), 4067–4081.
Blackhurst, J., Dunn, K. S., & Craighead, C. W. (2011). An empirically derived framework of global supply resiliency. *Journal of Business Logistics, 32*(4), 374–391.
Bode, C., Wagner, S. M., Petersen, K. J., & Ellram, L. M. (2011). Understanding responses to supply chain disruptions: Insights from information processing and resource dependence perspectives. *Academy of Management Journal, 54*(4), 833–856.
Brusset, X., & Teller, C. (2017). Supply chain capabilities, risks, and resilience. *International Journal of Production Economics, 184*, 59–68.
Casti, J. L. (1979). *Connectivity, complexity and catastrophe in large-scale systems*. Wiley.
Chen, L., & Miller-Hooks, E. (2012). Resilience: An indicator of recovery capability in intermodal freight transport. *Transportation Science, 46*(1), 109–123.
Craighead, C. W., Blackhurst, J., Rungtusanatham, M. J., & Handfield, R. B. (2007). The severity of supply chain disruptions: Design characteristics and mitigation capabilities. *Decision Sciences, 38*(1), 131–156.
Cui, R., Li, M., & Zhang, S. (2022). AI and procurement. *Manufacturing & Service Operations Management, 24*(2), 691–706.
Dolgui, A., & Ivanov, D. (2021). Ripple effect and supply chain disruption management: New trends and research directions. *International Journal of Production Research, 59*(1), 102–109.
Dubey, R., Gunasekaran, A., Childe, S. J., Wamba, S. F., Roubaud, D., & Foropon, C. (2021). Empirical investigation of data analytics capability and organizational flexibility as complements to supply chain resilience. *International Journal of Production Research, 59*(1), 110–128.
Elmachtoub, A. N., & Grigas, P. (2022). Smart "predict, then optimize". *Management Science, 68*(1), 9–26.
Fiksel, J., Polyviou, M., Croxton, K. L., & Pettit, T. J. (2015). From risk to resilience: Learning to deal with disruption. *MIT Sloan Management Review, 56*(2), 79–86.
Ghadge, A., Weiß, M., Caldwell, N. D., & Wilding, R. (2019). Managing cyber risk in supply chains: A review and research agenda. *Supply Chain Management: An International Journal, 25*(2), 223–240.

Ivanov, D. (2019). Disruption tails and revival policies: A simulation analysis of supply chain design and production-ordering systems in the recovery and post-disruption periods. *Computers and Industrial Engineering, 127*, 558–570.
Ivanov, D. (2020a). Predicting the impact of epidemic outbreaks on the global supply chains: A simulation-based analysis on the example of coronavirus (COVID-19/SARS-CoV-2) case. *Transportation Research: Part E, 136*, 101922.
Ivanov, D. (2020b). Supply chain simulation and optimization with anyLogistix.
Ivanov, D. (2020c). "A blessing in disguise" or "as if it wasn't hard enough already": Reciprocal and aggravate vulnerabilities in the supply chain. *International Journal of Production Research, 58*(11), 3252–3262.
Ivanov, D., & Das, A. (2020). Coronavirus (COVID-19/SARS-CoV-2) and supply chain resilience: A research note. *International Journal of Integrated Supply Management, 13*(1), 90–102.
Ivanov, D., & Dolgui, A. (2022). Stress testing supply chains and creating viable ecosystems. *Operations Management Research, 15*, 475–486.
Ivanov, D., Dolgui, A., & Sokolov, B. (2022). Cloud supply chain: Integrating Industry 4.0 and digital platforms in the "supply chain-as-a-service". *Transportation Research—Part E: Logistics and Transportation Review, 160*, 102676.
Klibi, W., Martel, A., & Guitouni, A. (2010). The design of robust value-creating supply chain networks: A critical review. *European Journal of Operational Research, 203*(2), 283–293.
Lücker, F., & Seifert, R. W. (2017). Building up resilience in a pharmaceutical supply chain through inventory, dual sourcing and agility capacity. *Omega, 73*, 114–124.
Lücker, F., Seifert, R. W., & Biçer, I. (2019). Roles of inventory and reserve capacity in mitigating supply chain disruption risk. *International Journal of Production Research, 57*(4), 1238–1249.
MacCarthy, B., & Ivanov, D. (2022). The digital supply chain—Emergence, concepts, definitions, and technologies. In B. MacCarthy & D. Ivanov (Eds.), *The digital supply chain* (pp. 3–14). Elsevier.
Macdonald, J. R., Zobel, C. W., Melnyk, S. A., & Griffis, S. E. (2018). Supply chain risk and resilience: Theory building through structured experiments and simulation. *International Journal of Production Research, 56*(12), 4337–4355.
Melnyk, S. A., Closs, D. J., Griffis, S. E., Zobel, C. W., & Macdonald, J. R. (2014). Understanding supply chain resilience. *Supply Chain Management Review, 18*(1), 34–41.
Mourtzis, D. (Ed.). (2022). *Design and operation of production networks for mass personalization in the era of cloud technology*. Elsevier.
Nair, A., & Vidal, J. M. (2011). Supply network topology and robustness against disruptions: An investigation using multiagent model. *International Journal of Production Research, 49*(5), 1391–1404.
Osadchiy, N., Gaur, V., & Seshadri, S. (2016). Systematic risk in supply chain networks. *Management Science, 62*(6), 1755–1777.
Papadopoulos, T., Gunasekaran, A., Dubey, R., Altay, N., Childe, S. J., & Wamba, S. F. (2017). The role of big data in explaining disaster resilience in supply chains for sustainability. *Journal of Cleaner Production, 142*(2), 1108–1118.
Pettit, J. T., Fiksel, J., & Croxton, L. K. (2010). Ensuring supply chain resilience: Development of a conceptual framework. *Journal of Business Logistics, 31*(1), 1–21.
Simchi-Levi, D., Schmidt, W., Wei, Y., Zhang, P. Y., Combs, K., Ge, Y., Gusikhin, O., Sander, M., & Zhang, D. (2015). Identifying risks and mitigating disruptions in the automotive supply chain. *Interfaces, 45*(5), 375–390.
Sodhi, M., & Tang, C. S. (2012). *Managing supply chain risk*. Springer.
Tomlin, B. (2006). On the value of mitigation and contingency strategies for managing supply chain disruption risks. *Management Science, 52*, 639–657.
Yoon, J., Talluri, S., Yildiz, H., & Ho, W. (2018). Models for supplier selection and risk mitigation: A holistic approach. *International Journal of Production Research, 56*(10), 3636–3661.
Zhang, G., MacCarthy, B., & Ivanov, D. (2022). The cloud, platforms, and digital twins—Enablers of the digital supply chain. In B. MacCarthy & D. Ivanov (Eds.), *The digital supply chain* (pp. 77–91). Elsevier.

# Supplementary Information




D. Ivanov, *Introduction to Supply Chain Analytics*, Classroom Companion: Business,
https://doi.org/10.1007/978-3-031-51241-4

# Index

## N



## O



## P



## Q



## R



## S



## T

GPSR Compliance
The European Union's (EU) General Product Safety Regulation (GPSR) is a set of rules that requires consumer products to be safe and our obligations to ensure this.

If you have any concerns about our products, you can contact us on

ProductSafety@springernature.com

In case Publisher is established outside the EU, the EU authorized representative is:

Springer Nature Customer Service Center GmbH
Europaplatz 3
69115 Heidelberg, Germany

www.ingramcontent.com/pod-product-compliance
Ingram Content Group UK Ltd.
Pitfield, Milton Keynes, MK11 3LW, UK
UKHW022210051026
15624UKWH00001B/97
* 9 7 8 3 0 3 1 5 1 2 4 0 7 *